城市设计科研与教育创新

——北京建筑大学2017年
国际城市设计联合工作营成果集

北京未来城市设计高精尖创新中心　组织编写

中国建筑工业出版社

图书在版编目（CIP）数据

城市设计科研与教育创新：北京建筑大学2017年国际城市设计联合工作营成果集/北京未来城市设计高精尖创新中心组织编写.—北京：中国建筑工业出版社，2018.11

ISBN 978-7-112-22758-7

Ⅰ.① 城… Ⅱ.① 北… Ⅲ.① 城市规划-建筑设计-研究-中国 Ⅳ.① TU984

中国版本图书馆CIP数据核字（2018）第226044号

责任编辑：田立平　李　璇
责任校对：李美娜

城市设计科研与教育创新——北京建筑大学2017年国际城市设计联合工作营成果集
北京未来城市设计高精尖创新中心　组织编写

*

中国建筑工业出版社出版、发行（北京海淀三里河路9号）
各地新华书店、建筑书店经销
北京锋尚制版有限公司制版
天津图文方嘉印刷有限公司印刷

*

开本：880×1230毫米　1/16　印张：8½　字数：203千字
2019年11月第一版　2019年11月第一次印刷
定价：110.00元
ISBN 978-7-112-22758-7
（32862）

版权所有　翻印必究
如有印装质量问题，可寄本社退换
（邮政编码 100037）

编委会
EDITORIAL BOARD

主　任：李雪华

副主任：吕小勇　李春青

委　员：（按姓氏笔画排序）
丁光辉　王　韬　王如欣　王秋爽　孙　瑶
任中琦　李　煜　李珊珊　肖　冰　张振威
郝石盟　俞天琦　徐加佳　蒋　蔚　穆　钧

前言
PREFACE

　　国际联合工作营是行业联合国际顶级专家开展城市设计创新研究的主要形式和通用模式。因城市规划、城市设计与建筑设计行业的特殊性，需要规划师、建筑师及科研团队成员通过现场调研、实地踏勘，在获取第一手资料基础上，探寻城市建设与发展的现存问题及原因，进而结合国际先进的实际案例和实践经验，依靠科学的研究路径与方法，在充分交流和研讨基础上，提出先进并具有较强创新性的方案成果。作为城市设计与管理领域高层次创新人才培养基地、高水平创新型城市设计国家级高端智库的北京未来城市设计高精尖创新中心，组织构建跨国界、跨文化、跨学科的国际城市设计联合工作营，旨在汇聚国内外跨领域、跨学科的知名专家共同参与城市设计与研究过程，以多元化的视角认知城市设计发展脉络，立足全球视野探索多样化的城市设计方法和实施路径；同时，力求搭建高水平的国际学术交流、国内外文化沟通桥梁，把握国际行业动态和前瞻专业思维，传播中国设计理念，输出中国建筑文化，实践未来国际城市设计人才创新培养模式，产出一批具有影响力的学术成果。

2017年，高精尖中心联合英国剑桥大学、美国密歇根大学、新加坡国立大学、美国迈阿密大学等6所全球知名高校，共同组织首届国际城市设计联合工作营，围绕城市更新设计、建筑可持续设计、交通规划管理、地下综合管廊等主题，进行联合学术调研、集中研讨、专项研究与规划设计，取得了丰硕的研究成果。本书作为高精尖中心2017国际城市设计联合工作营的研究成果集成，全面展示了高精尖中心国际联合教学、研讨和创新研究的实践经验，希冀为北京乃至中国城市设计领域发展提供可资借鉴的国际方案和创新思维。

目录
CONTENT

01

小镇风光·活力复兴
城市织补·磁力重塑
TOWN SCENERY-VIBRANT REVIVAL
URBAN WEAVING - MAGNETIC RESHAPING

英国剑桥大学联合工作营
Joint Workshop of the University of Cambridge

合作机构简介 Cooperative Institution Profile	2
团队简介 Team Profile	3
工作营任务书 Workshop Assignment	7
工作营日程 Workshop Schedule	9
团队成果简介与展示 Team Achievements and Presentation	11

02

健康城市设计：北京与纽约2017
HEALTHY URBAN DESIGN: BEIJING AND NEW YORK 2017

美国密歇根大学联合工作营
Joint Workshop of the University of Michigan

合作机构简介 Cooperative Institution Profile	20
团队简介 Team Profile	21
工作营任务书 Workshop Assignment	26
工作营日程 Workshop Schedule	28
团队成果简介与展示 Team Achievements and Presentation	31

03

城市设计视角下的日本都市空间
开发策略及实践调查研究
JAPANESE URBAN SPACE DEVELOPMENT STRATEGY AND SURVEY OF PRACTICE FROM THE PERSPECTIVE OF URBAN DESIGN

日本名城大学联合工作营
Joint Workshop of Meijo University

合作机构简介 Cooperative Institution Profile	48
团队简介 Team Profile	49
工作营任务书 Workshop Assignment	52
工作营日程 Workshop Schedule	53
团队成果简介与展示 Team Achievements and Presentation	55

04

可持续设计研究
RESEARCH ON SUSTAINABLE DESIGN

新加坡国立大学联合工作营
Joint Workshop of the National University of Singapore

合作机构简介 ———————— 62
Cooperative Institution Profile

团队简介 ———————————— 63
Team Profile

工作营任务书 ———————— 70
Workshop Assignment

工作营日程 ————————— 78
Workshop Schedule

团队成果简介与展示 ———— 80
Team Achievements and Presentation

05

生土建构实践
SOIL CONSTRUCTION PRACTICE

"土生土长"国际生土联合工作营
"Back to Earth" - International Earthen Joint Workshop

合作机构简介 ———————— 90
Cooperative Institution Profile

团队简介 ———————————— 91
Team Profile

工作营任务书 ———————— 97
Workshop Assignment

团队成果简介与展示 ———— 99
Team Achievements and Presentation

06

热带城市与建筑：
弹性海岸线城市设计
TROPICAL CITIES AND ARCHITECTURE: URBAN DESIGN FOR FLEXIBLE COASTAL LINES

美国迈阿密大学联合工作营
Joint Workshop of the University of Miami

合作机构简介 ———————— 106
Cooperative Institution Profile

团队简介 ———————————— 107
Team Profile

工作营任务书 ———————— 111
Workshop Assignment

工作营日程 ————————— 115
Workshop Schedule

团队成果简介与展示 ———— 116
Team Achievements and Presentation

TOWN SCENERY-VIBRANT REVIVAL
URBAN WEAVING - MAGNETIC RESHAPING

小镇风光·活力复兴
城市织补·磁力重塑

01

英国剑桥大学联合工作营

Joint Workshop of the University of Cambridge

合作机构简介
Cooperative Institution Profile

剑桥大学坐落于英国剑桥，是一所世界著名的公立研究型大学，采用书院联邦制，在许多领域拥有崇尚的学术地位及广泛的影响力，被公认为当今世界最顶尖的高等教育机构之一。

Located in Cambridge, UK, University of Cambridge is a world-renowned public research university. It is a federal university with superior academic status and wide influence in many fields, and is recognized as one of the best institutions of higher learning in the world today.

剑桥国际土地学院位于剑桥的圣约翰创新中心，专门为房地产、建筑环境和商业以及相关公共政策的专业和商业团体提供高质量的培训课程和考察研究。

Cambridge International Land Institute, based at the St John's Innovation Centre in Cambridge, specializes in high quality training courses and study tours for professional and business groups on real estate, built environment and business and related public policy.

该学院由皇家特许测量师学会于1993年成立，由剑桥大学土地经济系参与。其最初专注于英国房地产专业人士，之后扩大了其他专业课程，为专业公司提供房地产投资和金融服务；并增加了一系列国际课程，其中包括许多美国商学院、亚洲大学、专业协会和海外高级政府官员。

The Institute was founded in 1993 by the Royal Institution of Chartered Surveyors, with involvement of the Department of Land Economy, University of Cambridge. With an initial focus on UK property professionals, it has extended its reach with specialized courses in property investment and finance for professional firms, and added an international portfolio of courses, including many for American business schools, Asian universities, professional associations and senior oversea government official.

该学院以其卓越的课程内容和对交流访学安排的细节关注而享有盛誉。客座讲师都是来自英国和其他国家的知名国际专家。该学院所有课程的一个突出特点是与当代商业活动和当下实践直接相关。

The Institute has a strong reputation for the excellence of the content of its courses and the attention to detail in study tour arrangements. Guest Lecturers are drawn from leading international experts in the UK and elsewhere. A strong feature of all Institute programmes is the immediate relevance to the contemporary business activities and current practice.

团队简介
Team Profile

指导教师
Instructors

尼古拉斯·约翰·雷
Nicholas John Ray

利物浦大学客座教授、耶稣学院名誉会员；剑桥大学IDBE课程贡献者；曾任剑桥历史建筑集团创始人、董事；剑桥大学耶稣学院院长兼研究主任；剑桥大学建筑学讲师等职务。出版著作 *Rafael Moneo: Building, Teaching, Writing*, with Francisco Gonzalez de Canales, Yale University Press（2015）、*Philosophy of Architecture*, book with Christian Illies, Cambridge Architectural Press（2014）、"Philosophy of Architecture", chapter with Christian Illies in: *Philosophy of Technology and Engineering Sciences, Elsevier*（2009）等。

Visiting Professor at University of Liverpool, honorary member of the Jesus College; contributor to the IDBE Program at Cambridge University; founder and former director of the Cambridge Historic Buildings Group; dean and research director at the Jesus College of Cambridge; lecturer in Architecture, University of Cambridge; etc. Published works: Rafael Moneo: Building, Teaching, Writing, *with Francisco Gonzalez de Canales, Yale University Press (2015),* Philosophy of Architecture, *book with Christian Illies, Cambridge Architectural Press (2014), "Philosophy of Architecture", chapter with Christian Illies in:* Philosophy of Technology and Engineering Sciences, *Elsevier (2009), etc.*

杰米·利沃塞吉
Jamie Liversage

景观建筑师，曾参与香港、新加坡、迪拜和塞浦路斯等地区的酒店及度假村项目设计。曾与 Robert Holden 合作，在伦敦格林威治大学讲授景观建筑和园林设计课程。

Landscape architect who has worked on the design of projectss such as hotels and resorts in Hong Kong, Singapore, Dubai and Cyprus. He has worked with Robert Holden and teaches landscape architecture and garden design lessons at the University of Greenwich, London.

罗伯特库克曼
Robert Cookman

英国皇家农业大学国际地产教授。

Professor in International Real Estate at Royal Agricultural University, UK.

丁光辉
Ding Guanghui

北京建筑大学建筑与城市规划学院讲师，英国诺丁汉大学建筑学博士，香港城市大学博士后研究员。主要研究方向为当代中国的建筑实践（设计、评论、出版和展览等活动），研究论文发表在 *Architectural Research Quarterly, Journal of the Society of Architectural Historians, Habitat International* 等国际学术期刊上。已出版专著 *Constructing a Place of Critical Architecture in China: Intermediate Criticality in the Journal Time + Architecture*（2016）；与薛求理教授合著 *A History of Design Institutes in China: From Mao to Market*（2018）。

Lecturer at College of Architecture and Urban Planning, Beijing University of Civil Engineering and Architecture; PhD in Architecture, University of Nottingham, UK; postdoctoral researcher, City University of Hong Kong. His main research area is architectural practice in contemporary China (design, criticism, publication and exhibition etc). His research papers have been published in Architectural Research Quarterly, Journal of the Society of Architectural Historians, Habitat International *and other international academic journals. Published books include:* Constructing a Place of Critical Architecture in China: Intermediate Criticality in the Journal Time + Architecture *(Routledge, 2016); A History of Design Institutes in China: From Mao to Market (Routledge, 2018, co-author with Professor Xue Qiuli).*

李珊珊
Li Shanshan

北京建筑大学建筑与城市规划学院建筑系讲师，意大利都灵理工大学建筑学博士。主要研究方向为开放住宅的理论与实践，相关学术论文曾发表在 *Open House International* 等国内外学术期刊；已完成译著《支撑体与人民》（2018）。教学实践主要包括：《建筑设计（三）》（三年级秋季）、《建筑设计（六）》（四年级春季）等；主讲课程《居住建筑设计原理》（四年级春季）。

Lecturer at College of Architecture and Urban Planning, Beijing University of Civil Engineering and Architecture; PhD in Architecture at Polytechnic University of Turin, Italy. Her main research area is the theory and practice of open houses, and her relevant academic papers have been published in domestic and foreign academic journals such as Open House International. *She translated "Support and People" (2018). Her teaching practice includes: "Architectural Design (3)" (third year autumn), "Architectural Design (6)" (fourth year spring), etc.; and course "Residential Building Design Principles" (fourth year spring).*

团队学生
Team Students

甘振坤

遗产保护 – 博研 2016 级

Gan Zhenkun
Heritage Conservation - PhD 2016

张筱晶

建筑学 – 研 2015 级

Zhang Xiaojing
Architecture - Postgrad 2015

张梦宇

设计学 – 研 2015 级

Zhang Mengyu
Design - Postgrad 2015

陈旭

建筑学 – 研 2016 级

Chen Xu
Architecture - Postgrad 2016

李文博

城乡规划学 – 研 2016 级

Li Wenbo
Urban and Rural Planning - Postgrad 2016

张超

建筑学 – 研 2016 级

Zhang Chao
Architecture - Postgrad 2016

小镇风光·活力复兴　城市织补·磁力重塑

郝韵

城乡规划学－本 2013 级

Hao Yun
Urban and Rural Planning - Undergraduate 2013

李啟潍

建筑学－本 2013 级

Li Qiwei
Architecture - Undergraduate 2013

傅志颖

建筑学－本 2014 级

Fu Zhiying
Architecture -Undergraduate 2014

杨梅子

城乡规划学－本 2014 级

Yang Meizi
Urban and Rural Planning - Undergraduate 2014

工作照
Teams at work

工作营任务书
Workshop Assignment

项目背景介绍
Project background introduction

剑桥北站片区位于剑桥城市核心区的北部，也是当地绿色保护带的一部分。著名的康河穿越而过，自然景观优美，有高速铁路与伦敦市区连接，交通便利。场地周边环境条件十分复杂：既有的剑桥大学科技产业园和商业产业园位于北站的西部，传统的吉普赛人临时居住区在铁路以东、康河以西，场地周边还有一些工厂。由于河流和铁路的分割，场地的内部交通十分匮乏。

The Cambridge North is located in the northern part of the city and is part of the local green belt. The Cam River passes through the site, which is characterized with beautiful natural landscape. A high-speed railway connecting to London provides convenient transportation. However, the surrounding environmental conditions are very complicated: the existing Cambridge University Science and Technology Industrial Park and Commercial Industrial Park are located in the west of the North Station. The traditional Gypsies temporary residential area is in the east side of the railway, west of the Cam River, and there are some factories around the site. Due to the existence of river and railway, the internal traffic of the site is very inconvenient.

设计任务及要求
Design tasks and requirements

围绕剑桥北站开展综合片区的城市设计方案；
制定一个总体功能分布图；
重新规划片区的公共交通流线；
设计方案需要挖掘火车站附近的经济、社会和环境价值，平衡居住、办公、商业、休闲等功能需求，同时对康河景观带做出创造性回应。

Providing a conceptual urban design for the Cambridge North Station;
Developing an overall functional arrangement;
Re-planning the public transportation system;
The urban design plan needs to explore the economic, social and environmental values near the train station, balancing the functional needs of residence, office, commerce, leisure, etc., while making creative responses to the existing landscape.

工作营组织方式
Workshop organization

在中外双方教师的指导下,10 名学生被分为两个小组,分别负责提出一个设计概念,并在这个概念基础之上集体讨论,策划主要的功能安排,设计交通流线。经过充分的讨论交流和绘图,两组同学分别陈述各自的设计方案。之后,结合指导教师的点评,完善制作概念性城市设计方案。

Under the guidance of tutors, 10 students were divided into two groups, each of which was responsible for proposing a design concept through collective discussion, planning the main functional arrangements and designing transportation circulation. After a thorough discussion, the two groups of students presented their respective design plans to tutors. While incorporating the instructor's comments, two conceptual urban designs were produced.

工作营日程
Workshop Schedule

时间 Time		任务 Task
第一天 Day 1	上午 Morning	杰米·利沃塞吉和尼古拉斯·教授介绍工作营 Workshop Introduction by Jamie Liversedge and Professor Nicholas Ray 开场讲座：介绍剑桥 Introductory Lecture, on the expansion of Cambridge 集体考察迪顿沼泽地和剑桥北部工作营现场 Group Site Visit to Fen Ditton and North Cambridge Workshop Site "场所感"考察：理解场地环境、机遇与限制、特色与联系 Sense of Place visit to understand site context, opportunities and constraints, character and linkages
	下午 Afternoon	个人最初概念反馈及评图，形成两组 Individual first conceptual response session and pin up review. Two teams created 最初分组工作 First divided team session 各组最初概念方案深化及草图评图 Development of first conceptual proposal in teams and draft pin up review. 案例研究考查 Case Study reviews 技术研讨与临时讲座 Skills session & adhoc lectures
第二天 Day 2	上午 Morning	杰米·利沃塞吉介绍第二天工作营 Day 2 Workshop Introduction by Jamie Liversedge 专题讲座《剑桥建筑类型》 Introductory lecture, Cambridge building typologies 工作环节 Working Session
	下午 Afternoon	小组概念方案展示与细化评图 Presentation of team conceptual proposal and refined pin up & review 案例研究实地考察：剑桥阿科迪亚和阿博迪住宅区 Case Study Site Visit to Accordia and Abode developments, Cambridge 案例研究考查 Case Study reviews 技术研讨与临时讲座 Skills session & adhoc lectures

续表

时间 Time		任务 Task
第三天 Day 3	上午 Morning	杰米·利沃塞吉介绍第三天工作营 Day 3 Workshop Introduction by Jamie Liversedge 工作环节 Working Session
	下午 Afternoon	工作环节 Working Session 最终集中概念总图方案展示，最终小组项目评选 Presentation of final focused conceptual masterplan proposals and final group project selection 案例研究考查 Case Study reviews 技术研讨与讨论 Skills session & talks
第四天 Day 4	上午 Morning	杰米·利沃塞吉介绍第四天工作营 Day 4 Workshop Introduction by Jamie Liversedge 联合工作组环节 Combined Working Group Session 全组深化选中总图方案，最终格式及内容待定（各组联合） Full group development of selected masterplan proposal, final format and content to be confirmed (teams combined)
	下午 Afternoon	工作环节 Working Session
第五天 Day 5	上午 Morning	杰米·利沃塞吉介绍工作营 Workshop Introduction by Jamie Liversedge 工作环节 Working Session
	下午 Afternoon	工作环节 Working Session 最终总图展示 Final Presentation of developed masterplan 工作营项目总结 Workshop Project Summing Up

团队成果简介与展示
Team Achievements and Presentation

组一：小镇风光·活力复兴
Group 1: Town Scenery-Vibrant Revival

通过对场地的踏勘，我们发现横亘在场地范围内的铁路与河道将整个地块切分成了3部分，严重地阻碍了其间的交流与发展。协商后，我们一致认为可以引入桥来对整个场地进行织补，同时有效地协调其发展程度，通过对场地内的工厂等影响居住品质的地块进行功能置换等手段，尽最大可能地保留英国小镇的特色与韵味。除此，我们还依据其季节性的水位升降着力打造滨河景观，以改善其枯燥单调、道路通达性不良等现状。

By surveying the site, we found that the railway and the river channel through the area of the site cut the whole site into three, which seriously hindered the communication and development within. After consultation, we agreed that we could introduce bridges to weave the entire site, and thus effectively balance the level of its future development. We retain the characteristics and charming components of the British town as much as possible by means of functional replacement of plots that affect the quality of the property, such as factories on the site. In addition, we also build a riverfront landscape based on its seasonal water level change to improve its current status of monotony and poor road accessibility.

小镇风光·活力复兴　城市织补·磁力重塑

CHESTERTON FEN MASTERPLANNING WORKSHOP
TEAM-01

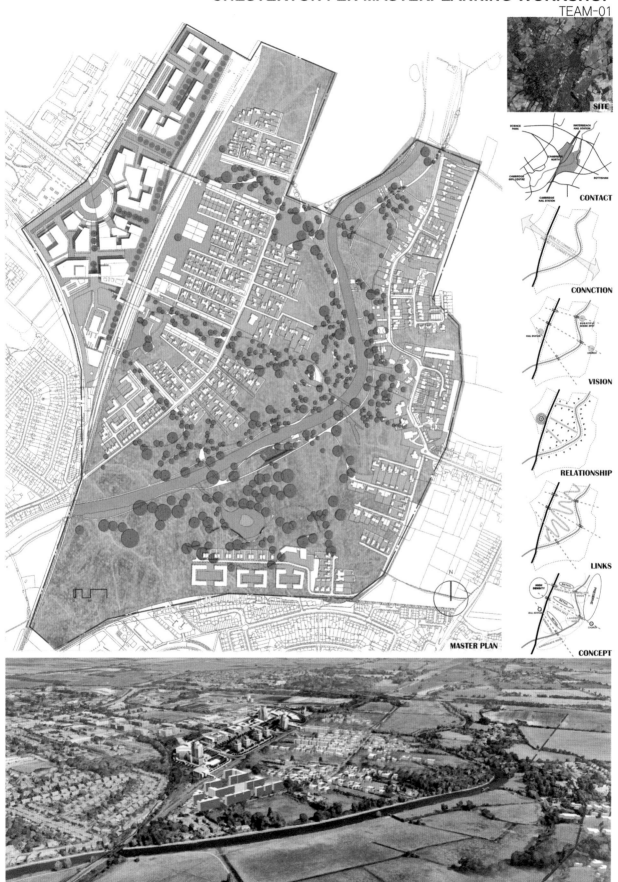

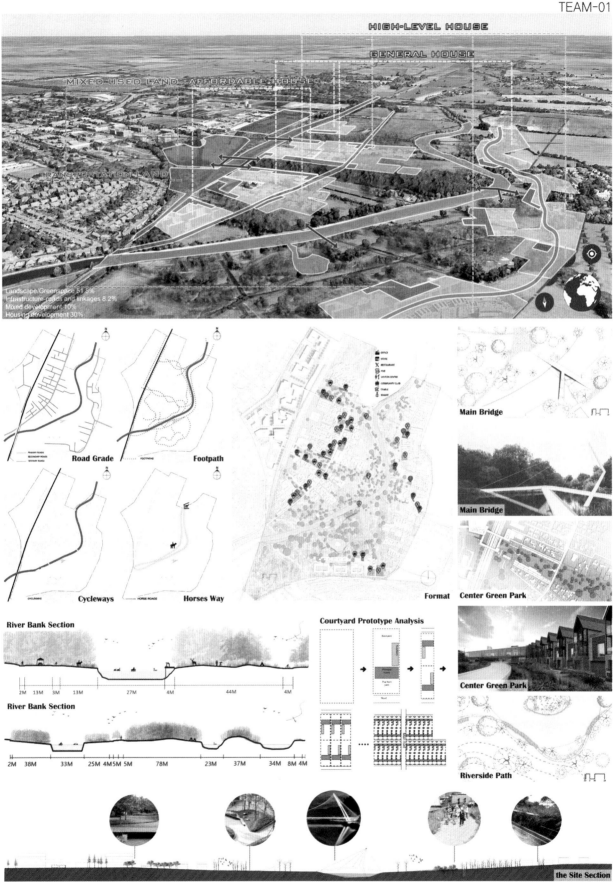

组二：城市织补·磁力重塑
Group 2: Urban Weaving - Magnetic Reshaping

方案场地选址为剑桥火车北站区，距离剑桥中心城区 3.8km。康河从场地中心穿流而过，两岸是环境优美的绿地和开发程度较低的住区。通往伦敦的铁轨自北向南布局在场地西侧，给区域带来交通便利的同时，一定程度也阻隔了轨道两侧的联系。与此同时，由于铁轨和河流的切割，场地中心片区形成了较为孤立的场所。

The site of the project is located in the Cambridge North Railway station area, 3.8km from downtown Cambridge. The Kang River flows through the center of the site, and the two banks contain beautiful green spaces and low-developed settlements. The railroads leading to London are laid out from the north to the south on the west side of the site, which brings convenience to the area and, to a certain extent, blocks the link across the track. At the same time, due to the segmentation of rails and rivers, the central area of the site formed a relatively isolated space.

针对场地存在的种种矛盾，方案提出了"城市织补·磁力重塑"的城市设计理念；通过强化联系、弃地再生、特色塑造三大策略，开展区域内的功能布局、交通流线、轨道站场、建筑景观的更新设计。方案旨在为高校学生、青年学者和本地居民创造出便捷的轨道上盖综合体、容积密度不尽相同的高品质住区，最终实现老城区的空间织补与活力复兴。

Faced with the various problems of the site, our design puts forward the urban design concept of "urban weaving and magnetic reshaping". Through three strategies of strengthening connections, brown field regeneration and feature shaping, we created an updated design for the region in terms of functional layout, traffic flow, railway station and architectural landscape. The project aims to create a convenient track assembly and high-quality living styles with different densities for college students, young scholars and local residents, and eventually realize space weaving and vitality revitalization for the old quarter.

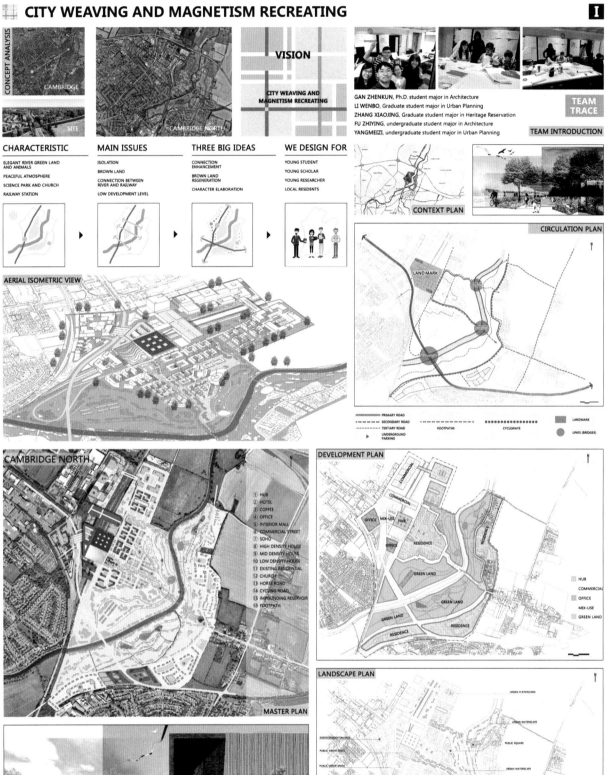

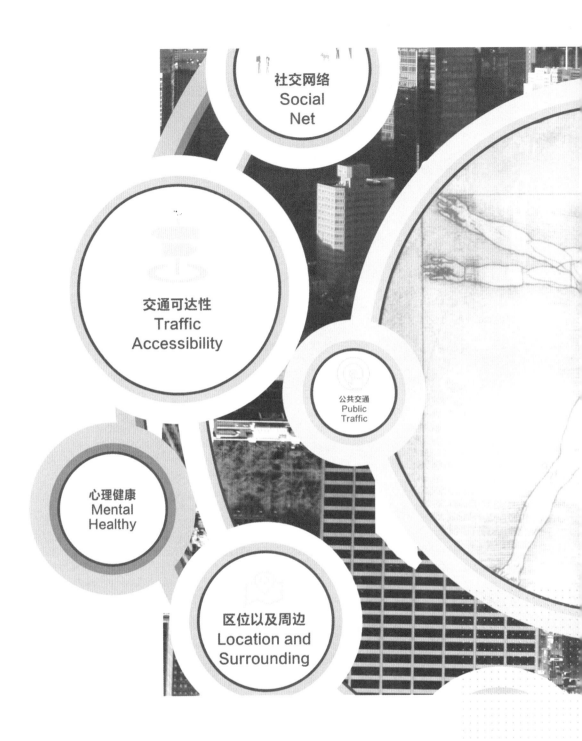

HEALTHY URBAN DESIGN:
BEIJING AND NEW YORK 2017

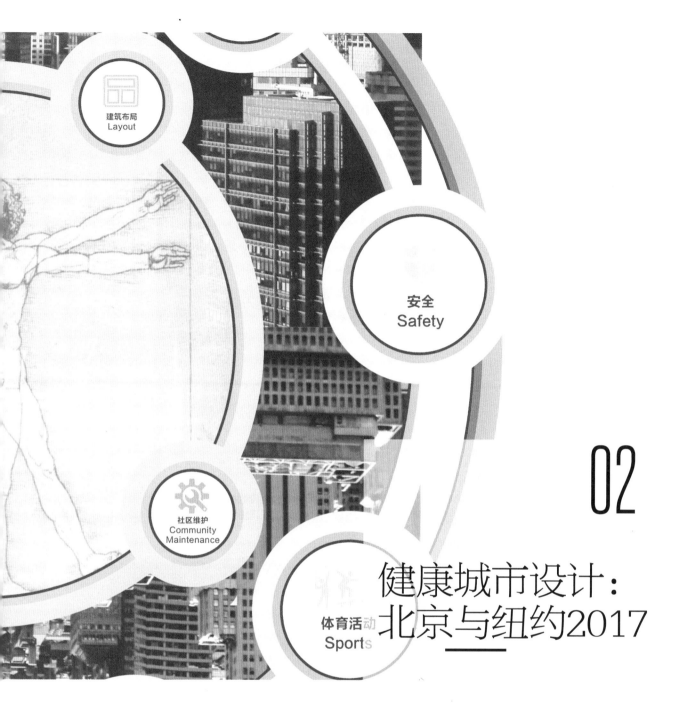

02

健康城市设计：北京与纽约2017

美国密歇根大学联合工作营

Joint Workshop of the University of Michigan

合作机构简介
Cooperative Institution Profile

密歇根大学（University of Michigan）创建于 1817 年，位于美国密歇根州，是美国历史最悠久的公立大学之一，被誉为"公立常春藤"和"公立大学的典范"，代表了美国公立大学的最高水平，在世界范围内享有盛誉。密歇根大学同时也是美国顶级学术联盟美国大学协会（Association of American Universities）的发起者之一。作为一所世界顶尖的综合研究型大学，密歇根大学在各学科领域中成就卓著并拥有巨大影响，学校的工程学院、医学院、商学院、法学院、文理学院、艺术学院等学院均位列全美前 15 名，超过 70% 的专业排名全美前 10 名。密歇根大学校友中包括 1 位美国总统、24 位诺贝尔奖得主、6 位图灵奖得主、8 位美国国家航空航天局宇航员、18 位普利策奖得主、25 名罗兹学术奖得主、30 多位各个大学的校长、上百位文艺娱乐界明星、上千位著名运动员以及不可计数的各行业的精英。

University of Michigan, founded in 1817, is located in Michigan State. It is one of the most long-standing public universities in the US, which is honored as the "Public Ivy" and "Model of Public Universities", representing the top of all American public universities. The internationally renowned UM is also a founding member of the top American academic union, Association of American Universities. As a globally leading comprehensive research university, UM stands out in various disciplines with remarkable achievement and huge influences. Its College of Engineering, Medical School, Law School, College of Literature, Science and Arts, Art & Design and many other rank among the top 15 in the US, and 70% of the disciplines enter the national top 10. UM's Alumni include one US president, 24 Nobel laureates, 6 Turing Award winners, 8 NASA astronauts, 18 Pulitzer winners, 25 Rhodes winners, presidents of over 30 universities, over a hundred art and entertainment celebrities, over a thousand famous sport stars and numerous elites in all professions.

团队简介
Team Profile

指导教师
Instructors

北京部分
Beijing

李煜
Li Yu

清华大学博士，北京建筑大学建筑与城市规划学院讲师，建筑系副主任，耶鲁大学访问学者（2013～2014），主要研究方向为城市设计、健康城市。

PhD at Tsinghua University, Lecturer at College of Architecture and Urban Planning, Beijing University of Civil Engineering and Architecture, deputy director of Department of Architecture, Visiting Scholar at Yale University (2013~2014). Research scope: urban design and healthy cities.

王韬
Wang Tao

浙江大学博士，北京建筑大学建筑与城市规划学院讲师，建筑系副主任，国家一级注册建筑师，主要研究方向为聚落人居环境更新设计。

PhD at Zhejiang University, Lecturer at College of Architecture and Urban Planning, Beijing University of Civil Engineering and Architecture, deputy director of Department of Architecture, National Grade I Registered Architect. Research scope: renewal design of human settlement.

纽约部分
New York

罗伊·斯特里克兰
Roy Strickland

密歇根大学斯特曼建筑与城市规划学院建筑学终身教授，城市设计研究生项目创始人、纽约城市设计论坛资深会员、美国《场所》杂志董事编委、建筑／城市设计教育家、北京建筑大学城市设计客座教授、北京未来城市设计高精尖创新中心学术委员会委员。

Tenured professor at Taubman College of Architecture and Urban Planning, University of Michigan, founder of the Urban Design graduate program, senior member of New York Urban Design Forum, member of the editorial advisory board of Places Journal, educator in architecture/urban design, guest professor at BUCEA, academic committee member of Beijing Advanced Innovation Center for Future Urban Design.

郝晓赛
Hao Xiaosai

清华大学博士，北京建筑大学建筑与城市规划学院副教授，建筑系主任（2013～），Center for Health Systems & Design, Texas A&M 访问学者（2017～2017）。主要研究领域为医院建筑设计，健康建筑等。

PhD at Tsinghua University, associate professor and director of Department of Architecture, College of Architecture and Urban Planning, BUCEA (2013~), visiting scholar at Center for Health Systems & Design, Texas A&M (2017~2017). Research scope: hospital design; health care building design, etc.

吴海燕
Wu Haiyan

同济大学博士，北京建筑大学土木与交通工程学院教授，从事道路设计和交通工程教学和研究28年。曾在英国利兹大学做访问学者。研究领域主要为城市交通政策和交通分析，熟悉北京城市交通演变过程。

PhD at Tongji University, professor at BUCEA, with 28 years of experience in teaching and research on Road Design and Traffic Engineering. Visiting scholar at University of Leeds. Research scope: urban transportation policies and traffic analysis, with good knowledge about the evolution of Beijing's urban transportation.

团队学生
Team students

单超

助教，建筑学 – 博研 2016 级

Shao Chao
Teaching Assisant, Architecture - PhD 2016

高朋辉

建筑学 – 研 2016 级

Gao Penghui
Architecture - Postgrad 2016

李慧东

建筑学 – 本 2014 级

Li Huidong
Architecture - Undergraduate 2014

罗棚

工业设计 – 本 2014 级

Luo Peng
Industrial Design - Undergraduate 2014

倪晨辉

建筑学 – 研 2016 级

Ni Chenhui
Architecture - Postgrad 2016

潘奕

建筑学 – 研 2016 级

Pan Yi
Architecture - Postgrad 2016

王海月

建筑学 - 研 2016 级

Wang Haiyue
Architecture - Postgrad 2016

王晴

建筑学 - 本 2013 级

Wang Qing
Architecture - Undergraduate 2013

杨雯琼

建筑学 - 研 2016 级

Yang Wenqiong
Architecture - Postgrad 2016

张泽宇

建筑学 - 研 2016 级

Zhang Zeyu
Architecture - Postgrad 2016

工作照
Teams at work

工作营任务书
Workshop Assignment

项目背景介绍
Project background introduction

本次工作营以"健康城市设计：北京与纽约 2017"为主题，在北京与纽约选择相关典型社区，开展健康城市数据采集与比较研究。师生赴北京、纽约两个世界级典型大城市的典型街区现场调研，就城市设计、健康城市等议题开展系列小讲座，由 Roy 教授全程指导工作营相关研究、评图。

This workshop takes "HEALTHY URBAN DESIGN: BEIJING &NEW YORK 2017" as the theme, chooses typical communities in Beijing and New York, and conducts healthy city data collection and comparative research. Our instructors and students will conduct site investigation and survey in some typical blocks of Beijing and New York, both world class typical metropolises, and organize a series of lectures on urban design and health city issues. Professor Roy will guide the related studies throughout the workshop and final review.

项目信息
Project information

北京部分
Beijing

北京健康社区 1：百万庄社区
Beijing healthy community 1: Baiwanzhuang Community

北京健康社区 2：新中街社区（东直门社区街道区域内，现状棚户区，紧邻东直门核心商圈）
Beijing healthy community 2: Xinzhongjie Community (in Dongzhimen community street area, currently a shanty town, next to the core business district of Dongzhimen.)

北京健康社区 3：新中西里社区
Beijing healthy community 3: Xinzhongxili Community

纽约部分
New York

纽约健康社区 1：纽约巴特雷公园城与曼哈顿下城
New York healthy community 1: Battery Park City and Lower Manhattan

纽约健康社区 2：扬·格尔百老汇大街城市更新与 59-14 街区
New York healthy community 2: Jahn Gehl Broadway re-design, 59th Street - 14th Street

纽约健康社区 3：高线公园及沿线
New York healthy community 3 : Highline

纽约健康社区 4：布鲁克林大桥公园
New York healthy community 4: Brooklyn Bridge Park

工作营组织方式
Workshop organization

学生每3人1组,共3组,完成北京3个社区和纽约3~4个社区的现场调研,并绘制相关图纸,完成对比研究。每组成果包括:
Three groups, each of three students, will conduct on-site investigation of three communities in Beijing and three to four in New York, and draw relevant drawings to complete the comparative study. Each set of results includes:

1．典型社区su分析模型（建筑、街道、公共空间、简单用地性质）。
1. Typical community analysis in SketchUp models (buildings, streets, public spaces, simple landuse properties) .

2．手绘调研表格（附件1）包含区内所有主要街道。
2. Hand drawn survey form (Appendix 1) which contains all the main streets in the district.

3．人员活动延时摄影视频（早、中、晚）。
3. Time-lapse photography of people's activity in the morning, afternoon, and evening.

4．Checklist 图示分析（附件2）。
4. Checklist diagram analysis (Appendix 2).

5．A1图纸（每个地块3张）。
5. A1-size drawings (3 pieces per plot).

工作营日程
Workshop Schedule

北京部分
Beijing

时间 Time		任务 Task
第一天 Day 1	上午 Morning	开营仪式 Opening Ceremony
	下午 Afternoon	专题讲座—健康与城市设计：介绍城市设计与健康城市基础内容 Lecture Health and Urban Design : Introduce basics of urban design and healthy city 开题：介绍任务、讲解调研地段、内容、分组 Thesis proposal: Introduce tasks, explain survey areas, content, and grouping
第二天 Day 2	上午 Morning	北京健康社区1调研（百万庄社区） SURVEY Beijing healthy community 1: Baiwanzhuang Community
	下午 Afternoon	图纸绘制，完成调研表格绘制、扫描、照片整理排版 Drawing, complete the survey form, scan, collect photos and prepare layout
第三天 Day 3	上午 Morning	专题讲座—城市肌理 场所 Lecture Urban fabric Loci
	下午 Afternoon	北京健康社区2调研（新中街社区） SURVEY Beijing healthy community 2: Xinzhongjie Community
第四天 Day 4	上午 Morning	专题讲座—健康城市设计——慢行交通 Lecture Healthy Urban Design—Non motorized traffic system
	下午 Afternoon	评图，分组汇报 Review,group reporting

续表

时间 Time		任务 Task
第五天 Day 5	上午 Morning	北京健康社区3调研（新中西里社区） SURVEY Beijing healthy community 3 : Xinzhongxili Community
	下午 Afternoon	图纸绘制，完成调研表格绘制、扫描、照片整理排版 Drawing, complete the survey form, scan, collect photos and prepare layout

纽约部分
New York

时间 Time		任务 Task
第一天 Day 1	上午 Morning	纽约调研任务布置，调研街区介绍 New York survey task arrangement, introduction of target blocks 专题讲座 Lecture
	下午 Afternoon	纽约健康社区1调研 Survey of New York healthy community 1
第二天 Day 2	上午 Morning	分组绘图、评图，完成调研表格等 Drawing and review in groups, complete the survey form etc
	下午 Afternoon	纽约健康社区2调研 Survey New York healthy community 2
第三天 Day 3	上午 Morning	双方师生会谈，交流活动介绍 Chinese and US instructors and students talk, introduction of communication activities 专题讲座A Lecture A
	下午 Afternoon	纽约健康社区3调研 Survey New York healthy community 3
第四天 Day 4	上午 Morning	专题讲座B Lecture B 分组绘图、评图，完成调研表格等 Drawing and review in groups, complete the survey form etc
	下午 Afternoon	纽约健康社区4调研 Survey New York healthy community 4
第五天 Day 5	上午 Morning	纽约健康社区5调研 Survey New York healthy community 5 分组绘图，完成调研表格等 Drawing in groups, complete the survey form etc
	下午 Afternoon	评图与交流 Review and discussion

调研表格设计
Design of Survey Form

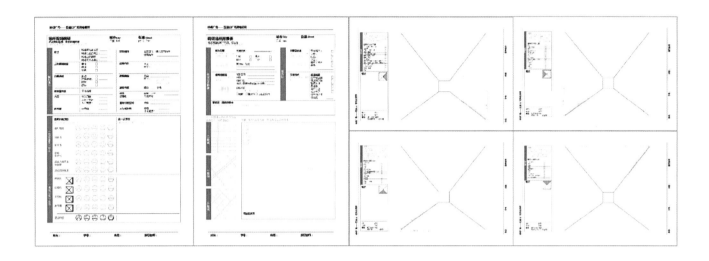

表格填写范例
Examples

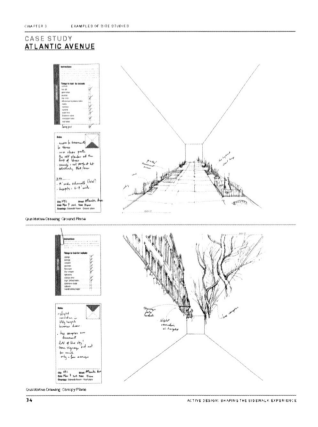
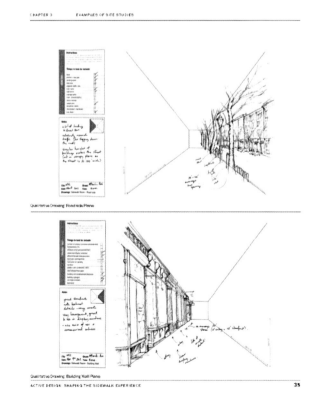

团队成果简介与展示
Team Achievements and Presentation

本次工作营以"健康城市设计：北京与纽约 2017"为主题，在北京与纽约选择相关典型社区，开展健康城市数据采集与比较研究。工作营前期在京调研北京旧城社区城市设计，后期赴美国纽约，在密歇根大学教授 Roy Strickland 教授的指导下，对纽约及周边城市社区的健康城市设计案例开展调研。师生就城市设计、健康城市等议题开展系列小讲座，学生每 3 人 1 组，共 3 组，完成北京 3 个社区和纽约 3～4 个社区的现场调研，并绘制相关图纸，完成对比研究，由 Roy 教授全程指导工作营相关研究、评图。3 个小组的项目图纸主要通过手绘记录和实景照片分析的方式，加入调研及居民访谈的内容，真实地呈现出被调研的街区生活空间的现状。之后 3 个小组合作汇总所得资料，将 7 个街区的类似项目进行统合，作客观地横向比较，以项目清单的方式进行直观表述。

The workshop, themed "Healthy Urban Design: Beijing and New York 2017", has chosen typical communities in Beijing and New York for Healthy Urban data acquisition and comparative studies. The team members first investigated the urban design elements in the old urban communities of Beijing. Later in New York, under the instruction of Prof Roy Strickland, the members visited New York and surrounding urban communities for Healthy Urban case studies. The instructors and students organized mini-lectures on urban design, Health Cities and other topics. Three teams, each consisting of three students, completed site investigation of three communities in Beijing and three to four in New York, and made drawings and comparative studies. Roy provided instruction throughout the process, and reviewed their results. The project drawings of the three teams have faithfully represented the current conditions of the living spaces in the communities by means of hand sketches and photo analysis, combined with records from site investigation and residents' interview. The three teams then put together all the documents, and synthesized similar projects of the seven communities for objective cross comparison, which is intuitively articulated in checklists.

通过观察评析中美两国城市典型城市空间对居民健康生活的各种影响，结合现场设计知识讲授，使学生深入理解并掌握健康城市设计的要义，学会城市设计的可实施方法。这次城市设计工作营不仅开拓了视野，提高了英语听说与交流能力，还在城市设计方法、设计实施方法、项目评析方法、甚至职业态度等方面多有斩获，对健康城市设计的内涵和外延有了深刻认知和实际感受，给未来的学习与工作打下了良好基础。

Through observation and analysis of the various impacts of typical Chinese and American urban spaces on residents' healthy life, the workshop, in combination of on-site teaching of design knowledge, has equipped the students with a deep understanding and grasp of the essentials of Health Urban Design, as well as a feasible methodology of urban design. The workshop not only enhanced the students' vision and English proficiency, but awarded them with methods of urban design, realization of design, and project analysis, even professional attitudes. Having facilitated a profound understanding and actual feelings of the connotation and extension of Healthy Urban Design, it paves the way for future study and work.

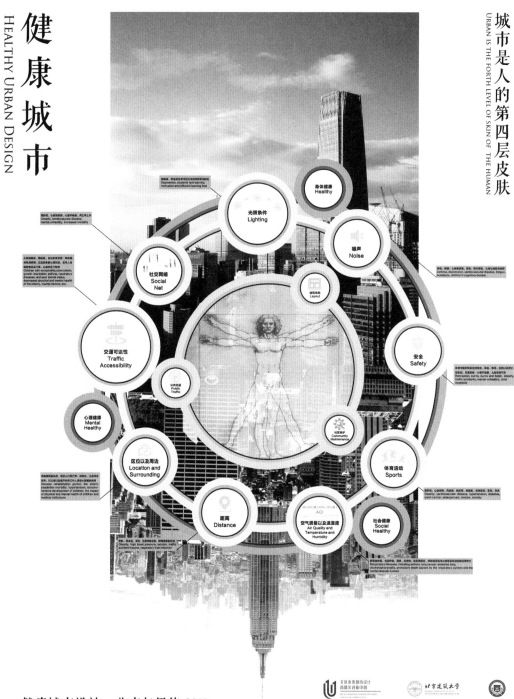

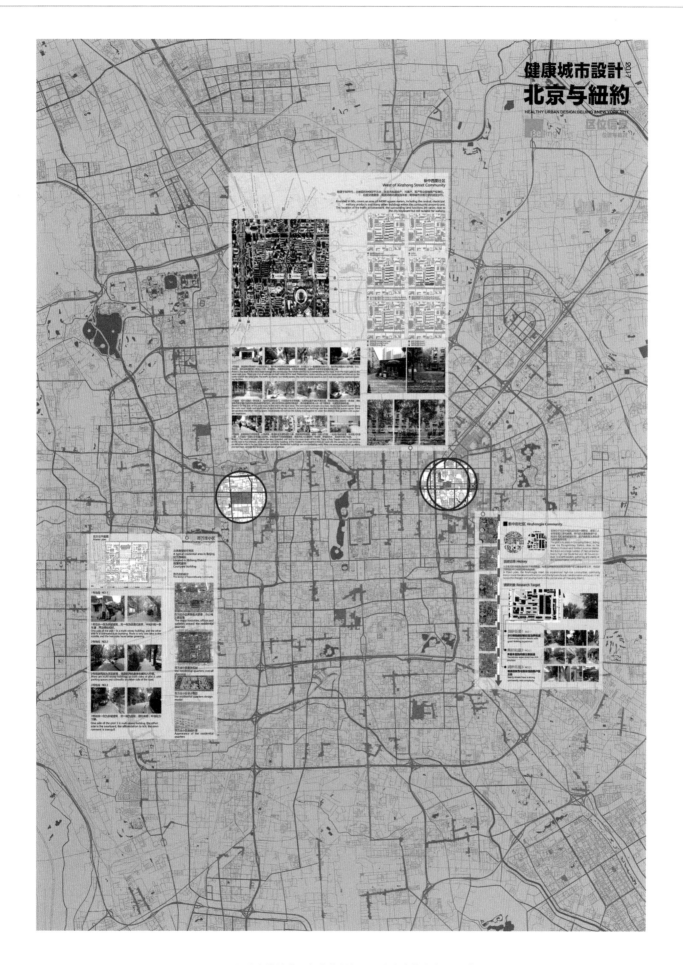

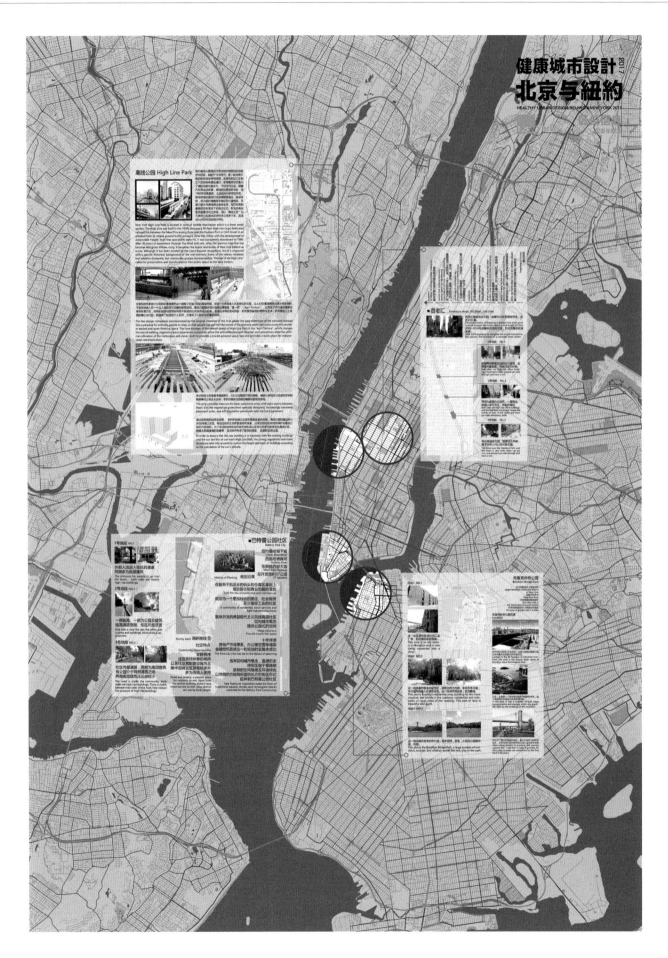

健康城市设计：北京与纽约 2017

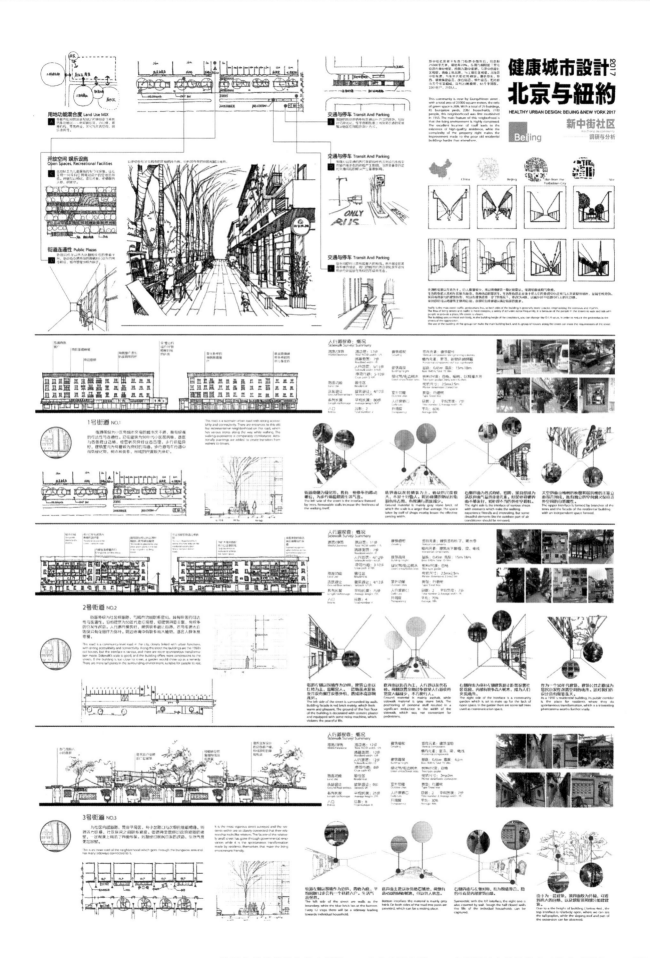

城市设计科研与教育创新——北京建筑大学 2017 年国际城市设计联合工作营成果集

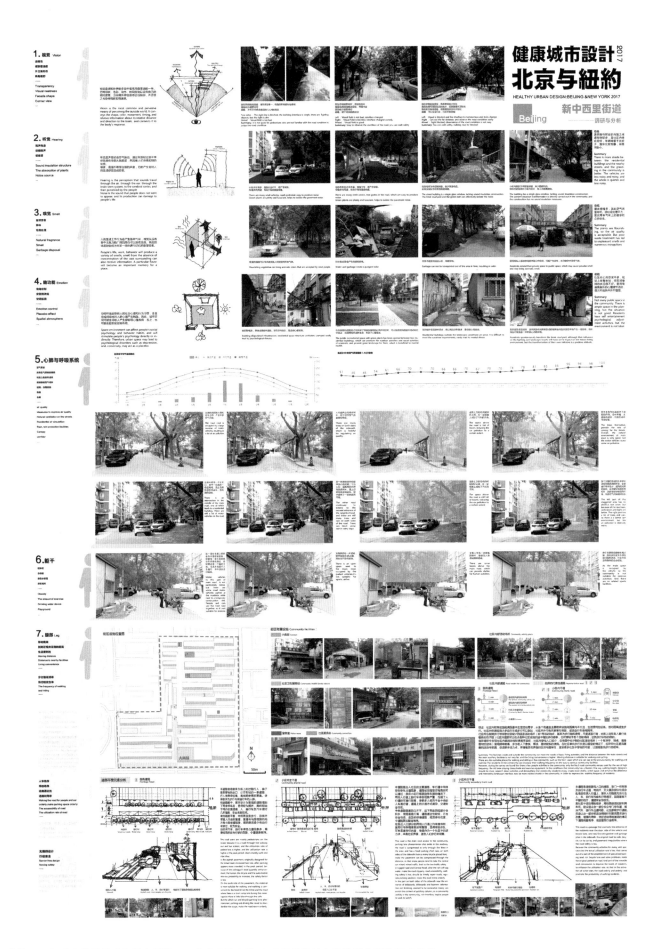

健康城市设计：北京与纽约 2017

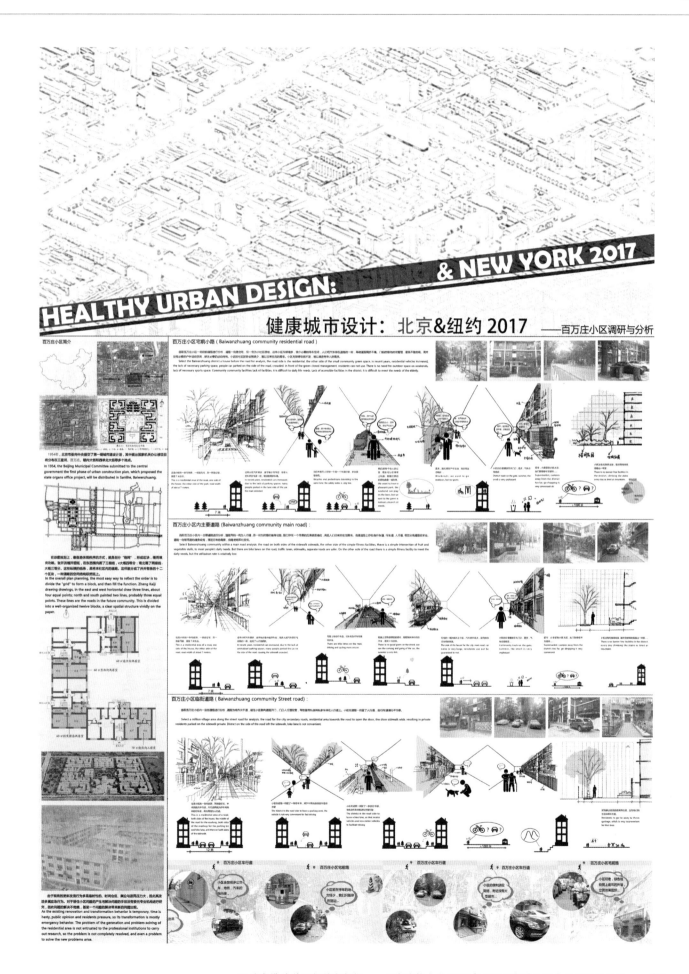

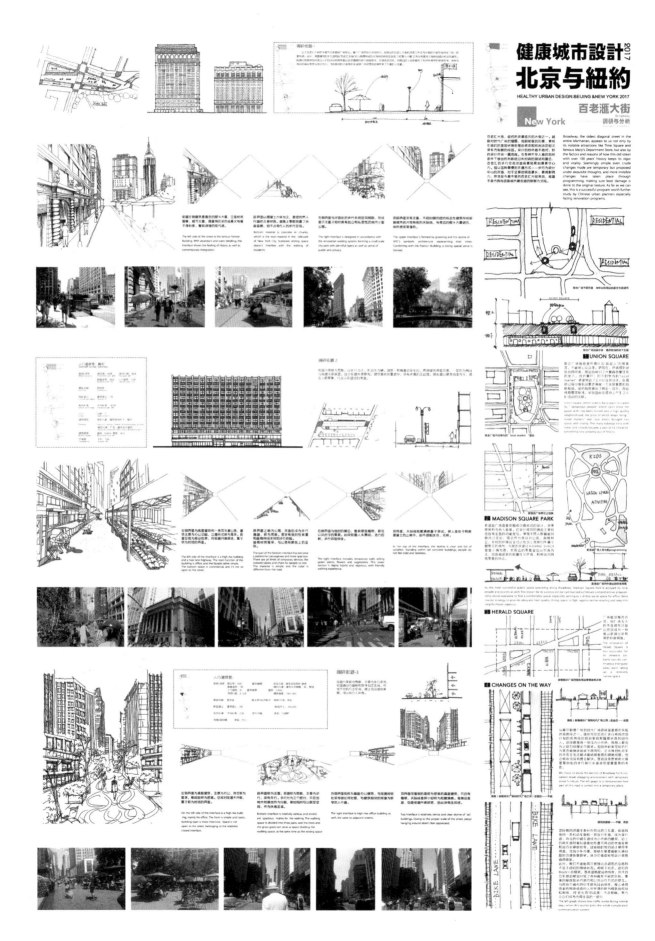

健康城市设计：北京与纽约 2017

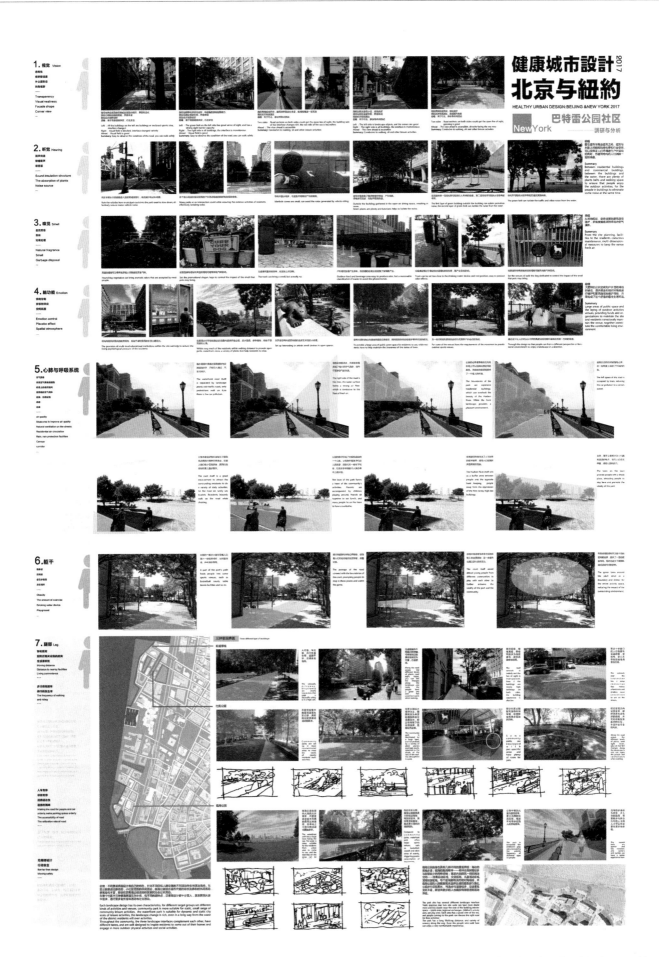

健康城市设计：北京与纽约 2017

城市设计项目清单 URBAN DESIGN CHECKLIST			四万庄社区 Baiwanzhuang Community	新中街社区 Xinzhongjie Community	新中街社区 West of Xinzhong Street Community	炮台公园城区 Battery Park City	百老汇大街城市规划 Broadway re-design, 59th Street - 14th Street	高线公园及街区 High Line Park	布鲁克林大桥公园 Brooklyn Bridge Park	总结 conclusion
公共广场 PUBLIC PLAZAS		创造有吸引力的广场空间并妥善维护。 Create attractive plaza spaces that are well-maintained.						✓		在北京的调研中，四万庄社区和新中街社区，新中街西的小区也设置了广场。在中国，新中社区和西街西的小区也设置了广场，可以通过自行车。Xinzhongjie Community and West of Xinzhong Street Community also have squares which are good. 纽约的巴特利公园和布鲁克林大桥公园也设置了广场，有广场维护得好，开设交通方便，可以广场中可以是小型人员休闲的场所，提供良好的休息和娱乐场所。In New York, the Battery Park and Brooklyn Bridge Park also set the square, they had well maintained, convenient transportation, convenient facilities. The square providing a good rest and entertainment venues for people within the district. 高线公园没有大广场，由于其线性布局的原因，因此，没有大广场。There are many visitors in the High Line Park. There are no big squares because it has been reconstruction from railway line. 百老汇大街沿线设有大型广场，为游客、上班族人提供了休息的场所。Along the Broadway street there are large squares, providing a place to rest for tourists and workers.
		在人流密集的行街旁边设置广场。 Locate public plazas along popular pedestrian streets.				✓		✓		
		在公交站点附近设置广场。 Locate plazas near transit stops.								
		使广场可以通行自行车。 Make plazas accessible to bicyclists.			✓					
		设置与人行道齐平的广场。 Create plazas that are level with the sidewalks.	✓		✓	✓				
		设计可容纳多功能使用的广场。 Design plazas that allow for diverse functions.			✓	✓				
		设计可在各种气候状况下使用的广场。 Design plazas to accommodate use in a variety of weather conditions.			✓	✓		✓		
		寻求可社会团体的合作关系以维护和运营广场。 Seek partnerships with community groups to maintain and program plazas.		✓		✓				
杂货店和生鲜食品进入入口 GROCERY STORES AND FRESH PRODUCE ACCESS		在所有居民区步行距离内提供全服务杂货店的食品店。 Develop full-service grocery stores within walking distance in all residential neighborhoods.			✓		✓			在居民聚集地附近提供各种日常的食品店住的，可以减少购物时间出行难度，避免长途的开车到外地区。In the vicinity of residents settlements to provide easy access to the miscellaneous department store, could encourage residents to travel short distance, to avoid long-distance driving procurement. 北京区三个地块都在中国是比区的规划和社区建设的好的原因，四个地块除了西于街之外，其他几个地块周边都有比较完备的商业购物建筑，居民的自我更进和卫生环境都较好。Beijing area three plots in addition to the West of Xinzhong Street community outside the planning and construction of a good commercial shopping building, the residents are more self-built, self-improvement of small food stores. Followed by the health environment and traffic problems. 纽约四个地块公园型的较特殊，其余地块都或多或少有一些百货大型的规划设置，继续了其小邻区、多功能的规划方式。New York in addition to the High Line Park with a special type, the rest of the plots are more or less included in the miscellaneous department store planning settings, continued its small neighborhood, multi-functional planning style.
		引入农贸市场作为食品店的补充。 Introduce farmer's markets as a complement to grocery stores.		✓		✓			✓	
		在人口密集的地区、食杂店和农贸市场之间提供人行和自行车道。 Provide safe walking and bicycle paths between densely populated areas and grocery stores and farmer's market sites.	✓		✓	✓				
		设计杂货店的布局和停车，使行人、自行车、汽车和装车能够安全便利地通行。提供停车场等自助设施如自行车停车和饮水喷泉。 Design grocery store layouts and parking to accommodate pedestrian, cyclist, automobiles, and loading trucks safely and conveniently. Provide infrastructure such as bicycle parking a drinking fountains.				✓				
街道连通性 STREET CONNECTIVITY		在大尺度的开发区中，设计有良好连接的道路和人行道，并保持相对小的街区尺度。 In large-scale developments. Design well-connected streets with sidewalks and keep block sizes relatively small.			✓			✓		北京的三个地块都具有一定年代的时间地方，街道节约比较小，也基本没有有改造的工程。Beijing's three plots are residential areas of a certain age, the size of the street block is relatively small, and basically no construction works. The streets connectivity is good. 不足之处在于，不断之处其自行车道往往不便被停，普通系统有效独立人行道（或者人行道被用作他用。The disadvantage is that for pedestrian system is not perfect enough, and there are no independent sidewalks in many places (or sidewalks are occupied by debris). 纽约的四个地块中，高线公园是比较独特的一个，它是一个线性布局系统的景区，其他三个地块每个都有特色。步行连通性好，社区舒适性好，连续、立体、舒适的人行道。Among the four plots in New York, the High Line is the most unique one. It is a linear park with a walking system. The rest of the three plots have better connectivity, comfortable living communities, and continuous, dimensional, comfortable sidewalks.
		如果目前建筑工地上人行道和街道的连通性较差，则通过现有的街区连接行人通道。 Where current connectivity of sidewalks and streets on a building site is poor, provide pedestrian paths through existing blocks.			✓	✓		✓		
		避免使用天桥或地道这些迫使行人改变楼层的设施。 Avoid creating pedestrian over- and underpasses that force walkers to change levels.								
		在城市道路的死胡同当中也为人行和自行车提供连接。 Maintain dedicated pedestrian and bicycle paths on dead-end streets to provide access even where cars cannot pass.						✓		
		在繁忙步行人流街道中减少上端加街距动车过街驶驶的路径。 Minimize addition of mid-block vehicular curb cuts on streets with heavy foot traffic.						✓		
		设计车道、坡道、减少车辆和行人之间的接触。 Design vehicular driveways and ramps to minimize contact between cars and pedestrians.			✓	✓	✓	✓		
交通稳化 TRAFFIC CALMING		设计道路的宽度为最小的实用和最少车道数。 Design roads to be minimum width and to have the minimum number of lanes practical.	✓	✓				✓		北京和纽约的街道路段设计有不同的特色。北京和纽约的街区、社区道路设计发展基本吻合，都能够满足行人和车辆交通需要。In Beijing and New York, the street road design has different characteristics. The width of road in Beijing's and New York's community is basically suitable. The road are able to meet the pedestrian and vehicle traffic demand. 纽约的街区配备有人行道扩展带、路障及中央、降低速坡。纽约的街区道路配有小型交通圈、"行人礼让"标志的设置比北京要全面许多。New York's roads are equipped with curb extensions, medians and raised speed reducers. In the streets of Beijing, the design of curb extensions, medians and raised speed reducers are insufficient. Other road design such as Traffic roundabouts into traffic circles and "Yield to Pedestrian" signs in New York is also more comprehensive than Beijing. 总而言之，纽约的道路交通稳化设计比北京要好。In a word, the traffic calming of the road in New York is better than those in Beijing.
		包含交通稳化措施，设置道路区扩展带、设置路障及中央、降低速坡等。 Incorporate traffic calming street additions such as curb extensions, medians, and raised speed reducers.				✓		✓		
		考虑其他合适的设计措施 Consider other physical design measures where appropriate.	使水平转向，譬如弯曲路面等，使垂直转向，譬如路面收起或交叉口。 Horizontal deflections such as curved roadway alignments vertical deflections such as raised intersections or crossings			✓	✓				
			交通分流、环形交叉、小型交通圈。 Traffic diverters, roundabouts, and mini-traffic circles.		✓	✓	✓	✓	✓		
			调整交通信号灯并保护停车左转弯相位设计。 Signal phasing plan with a protected left-turning lag phase								
			"礼让行人" 标志 "Yield to Pedestrian" signs				✓		✓		
			避免滑车道和宽路缘线半径。 Avoidance of slip lanes and wide curb radii				✓				
编程街景 PROGRAM-MING STREET-SCAPES		包含临时的和永久性的公共艺术作品的设置。 Incorporate temporary and permanent public art installations into the streetscape.				✓	✓	✓	✓	北京的三个社区都是住宅社区小区，主要解决居民居住的生活问题，所以对于街道的景观相对不会有过多的考虑。即使问题还存在一定。Beijing's three communities are relatively real residential areas, there are still many living environment problems, so for the street view is basically no consideration. Even though some communities have many shops on the street, there is a lack of standardized management. Although some shops strengthen Street activities, they occupy the space of the sidewalk and affect the continuity of the sidewalk. 纽约的四个地块，由于其用地性质或区的不同，街景设计差异较大。炮台公园城和布鲁克林大桥公园有多功能活动场所，而公园与街道距离较远，公共空间的艺术氛围设施，如河边广场。百老汇和高线公园街区以单一使用功能为主，无居住区和完善的公共艺术设施。New York's four plots, because of its different nature of land, streetscape difference is relatively large. Battery Park City and Brooklyn Bridge park are multifunctional activities, but the park is far from the streets, and there are more artistic facilities in the urban public space, such as the riverside or the square. The functional of the High Line and Broadway blocks are single, with no residential areas and complete public art facilities.
		组织面向行人的项目，如慈善徒步行和车辆街道封路，可以为行走和骑自行车者提供使用路径的活动。 Organize pedestrian-oriented programs, such as charity walks and vehicular street closures, that make wide avenues available for walking and bicycling.		✓	✓	✓	✓		✓	
		增加户外咖啡馆的数目，以加强街道活动。 Increase the number of outdoor cafes to enhance street activity.			✓	✓			✓	

健康城市设计：北京与纽约 2017

城市设计科研与教育创新——北京建筑大学 2017 年国际城市设计联合工作营成果集

JAPANESE URBAN SPACE DEVELOPMENT STRATEGY AND SURVEY OF PRACTICE FROM THE PERSPECTIVE OF URBAN DESIGN

城市设计视角下的日本都市空间开发策略及实践调查研究

日本名城大学联合工作营

Joint Workshop of Meijo University

合作机构简介
Cooperative Institution Profile

日本名城大学是一所位于日本爱知县名古屋市的私立大学，是日本中部地区首屈一指的文理型综合大学。学校的前身为日本物理学者田中寿一先生于 1926 年创立的名古屋高等理工科学校，于 1949 年改为新制大学。

Meijo University is a private university in Nagoya, Aichi Prefecture, Japan, known as a top-level comprehensive university of Arts and Sciences in Chubu region (Central Japan). The university grew out of the Nagoya Science and Technology Course, founded by Japanese physicist Mr. Juichi Tanaka in 1926, which was later chartered as a university in 1949.

建校以来，名城大学与名古屋大学等日本中部地区著名学府一直保持着十分密切的学术交往和人员往来关系。2014 年，名城大学教授赤崎勇和前教授天野浩获得了当年的诺贝尔物理学奖。

Since its establishment, Meijo University has maintained close academic communication and staff exchange with Nagoya University and other renowned higher education institutions in Chubu region. In 2014, Professor Isamu Akasaki and ex-Professor Amano Hiroshi, both from Meijo University, won the honor of Nobel Prize in Physics.

团队简介
Team Profile

指导教师
Instructors

张振威
Zhang Zhenwei

北京建筑大学建筑与城市规划学院讲师。毕业于清华大学建筑学院，在清华大学法学院从事博士后研究工作，美国德州大学奥斯汀分校、意大利都灵大学访问学者。主要研究方向有城市与景观法律与政策，景观生态规划、景观规划设计。主持国家自然科学基金1项，北京市教委社科重点项目1项，参加国家社科重大、国家自然科学基金5项。在《国际城市规划》《清华法治论衡》《中国园林》《风景园林》等刊物上发表论文10余篇。

lecturer at College of Architecture and Urban Planning, Beijing University of Civil Engineering and Architecture. After graduating from School of Architecture, Tsinghua University, he diverted to post-doc research at School of Law, Tsinghua University, and travelled to University of Texas at Austin and University of Turin as visiting scholar. His research interests include Law and Policy for City and Landscape, Landscape Ecological Planning, and Landscape Planning and Design. He has conducted one National Natural Science Fund (NNSF) project, and one Key Social Science Project for Beijing Municipal Commission of Education, and contributed to altogether five National Social Science Fund (NSSF) and NNSF projects. He has published over 10 papers on Urban Planning International, Tsinghua Journal of Rule of Law, Chinese Landscape Architecture, and Landscape Architecture etc.

福岛 茂
Sigeru FUKUSHIMA

日本名城大学副校长，日本名城大学都市情报学部城市信息学教授。
Vice President of Meijo University, professor of urban informatics, division of urban science, Meijo University, Japan.

1992年3月，于东京大学工学系博士毕业。
In March 1992, he graduated from the engineering department of Tokyo university.

现为日本城市规划学会、日本计划行政学会、日本经济地理学会、日本环境共生学会、日本远程办公学会会员。
He is now a member of Japan urban planning society, Japan planning administration society, Japan economic geography society, Japan environmental symbiosis society and Japan telecommuting society.

长期从事城市空间规划的理论研究和相关实践，主要研究领域为城市规划、国土和地区规划、建筑规划、住宅政策、建筑环境设备等，研究成果丰硕，先后发表学术著作10余册，学术论文60余篇，国际学会报告10余篇；积极组织并参与日本中小学教育志愿活动；为日本以及亚洲城市规划研究、教育做出了巨大贡献。
He has been engaged in the theoretical research and related practice of urban space planning for a long time, and his main research fields are urban planning, land and regional planning, architectural planning, housing policy, building environment and equipment, etc. He has obtained fruitful research results, and has published more than 10 volumes of academic works, more than 60 academic papers, and more than 10 reports of international society. He actively organize and participate in Japanese primary and secondary education volunteer activities;He has made great contributions to urban planning research and education in Japan.

团队学生
Team Students

李超

建筑学 - 研 2016 级

Li Chao
Architecture - Postgrad 2016

曹政

城乡规划学 - 研 2016 级

Cao Zheng
Urban and Rural Planning - Postgrad 2016

张伟锋

建筑学 - 研 2016 级

Zhang Weifeng
Architecture - Postgrad 2016

杨倩

建筑学 - 研 2016 级

Yang Qian
Architecture - Postgrad 2016

金璇

建筑学 - 研 2015 级

Jin Xuan
Architecture - Postgrad 2015

工作照
Teams at work

工作营任务书
Workshop Assignment

项目背景介绍
Project background introduction

日本的城市设计从注重"硬件"开始,到创造出符合日本国情、注重"软件"的城市设计理论与方法。

Urban design in Japan starts from an emphasis on the "hardware", which is followed by the creation of an urban design theory and methodology that focus on the "software" and that respects the national conditions in Japan.

北京未来城市设计高精尖创新中心与日本名城大学开展暑期联合工作营。我方由1名教师、5名学生组成团队,赴日本名古屋开展合作研究与研讨。

Beijing Advanced Innovation Center for Future Urban Design is organizing a joint summer workshop with Meijo University. The visiting team, which consists of one instructors and five students, will travel to Nagoya for collaborative research and discussion.

设计任务及要求
Design tasks and requirements

本工作营主题为"城市设计视角下的日本都市空间开发策略及实践调查研究",调研集中在日本式的"城市开发设计理论与实践"在日本城市发展变化中的几个问题的研究:老城的不同地段的开发及保护策略,新城建设的策略,社区改良,城市设计的法律地位等。基于这些方面的调查研究,本工作营将使学生对于日本都市的空间构思及设计实践有深入的认识及感受。

The joint summer workshop is themed "Japanese Urban Space Development Strategy and Survey of Practice from the Perspective of Urban Design". The survey is focused on the following issues of the Japanese "Design Theory and Practice for Urban Development" in the evolution of Japanese cities: development and conservation strategy of different sites in the old city, strategy of new urban development, community improvement, legal status of urban design etc. Based on these surveys, the workshop seeks to equip students with an in-depth understanding and experience of Japanese urban spatial conception and design practice.

工作营日程
Workshop Schedule

时间 Time		任务 Task
第一天 Day 1	上午 Morning	开营仪式 Workshop Opening Ceremony
	下午 Afternoon	专题讲座：日本都市开发调研——基于文化比较视角 Lecture : A Survey of Japanese Urban Development—from a Cultural Comparative Perspective
第二天 Day 2	上午 Morning	专题研讨：日本土木学发展历程——以名古屋为例 Seminar :A History of Civil Engineering in Japan—the Case of Nagoya
	下午 Afternoon	名城大学校园规划调研；名古屋规划馆调研 Survey of Campus Planning in Meijo University ;Visit to Nagoya Planning Pavilion
第三天 Day 3	上午 Morning	专题研讨：日本环境建设 Seminar :Environmental Development in Japan
	下午 Afternoon	工作营：名古屋城市设计与公共空间调研 Workshop :Survey of Nagoya Urban Design and Public Spaces
第四天 Day 4	上午 Morning	专题研讨：日本建筑历史 Seminar :Japanese Architectural History
	下午 Afternoon	工作营：名古屋城市设计与公共空间调研 Workshop :Survey of Nagoya Urban Design and Public Spaces
第五天 Day 5	上午 Morning	专题研讨：日本语言与文化学习 Seminar :Japanese Language and Culture Studies
	下午 Afternoon	专题研讨与工作营：日本都市开发历史研究 Seminar and Workshop :A Research of the History of Japanese Urban Development
第六天 Day 6	上午 Morning	工作营：日本都市开发与公共空间 Workshop :Japanese Urban Development and Public Spaces
	下午 Afternoon	工作营：日本都市开发制度研究 Workshop :A Research of Japanese Urban Development Institute

续表

时间 Time		任务 Task
第七天 Day 7	上午 Morning	专题研讨：日本都市规划体系 Seminar: the Japanese Urban Development System
	下午 Afternoon	工作营：日本都市规划与公共空间 Workshop: Japanese Urban Planning and Public Spaces
第八天 Day 8	上午 Morning	琵琶湖景观保护调研；京都城市设计与公共空间营造调研 Survey of Lake Biwa Landscape Conservation; Survey of Kyoto Urban Design and Public Space Development
	下午 Afternoon	工作营：日本城市景观保护研究 Workshop: Japanese Urban Landscape Conservation Studies
第九天 Day 9	上午 Morning	奈良城市设计与公共空间调研 Survey of Nara Urban Design and Public Space
	下午 Afternoon	工作营：日本城市公共空间比较研究 Workshop: A Comparative Study of Japanese Urban Public Space

团队成果简介与展示
Team Achievements and Presentation

本工作营以"城市设计视角下的日本都市空间开发策略及实践调查研究"为主题。在日本名城大学的周密安排下，工作营师生先后对日本都市规划体系、日本都市开发历史、日本都市开发制度，奈良古城保护、名古屋旧城保护、琵琶湖景观保护、名古屋大学新校区校园规划设计、建筑抗震、城市防灾减灾等方面进行考察调研与学习，对日本公共空间（包括旧城保护）形成机制机理的关键要素与驱动因子有了深入认识。在与名城大学共同开展的为期 2 天的团队 workshop 中，两校学生分为混编 4 组，以名古屋为例，对城市公共空间设计导则的原则、内容、方法以及其他关键要素进行了概念性建构，从名古屋城市开发、校园建设、住宅设计、遗产保护、生态环境保育五个方面进一步研究日本城市设计与城市空间，从不同角度理解建筑与城市的关系。

This workshop focus on the theme "Japanese Urban Space Development Strategy and Survey of Practice from the Perspective of Urban Design". Under the complete and thorough arrangement by Meijo University in Japan, teachers and students from the workshop have successively carried out urban planning system, urban development history, urban development system, Nara ancient city protection, Nagoya old city protection, Biwa Lake landscape protection, Nagoya University new campus planning and design, building earthquake resistance, urban disaster prevention and mitigation, etc. Through investigation, research and study, we have a deep understanding of the key elements and driving factors of the formation mechanism of Japanese public space (including old city protection). In a two-day team work process with Meijo University, the students of the two universities were divided into four mixed groups. Taking Nagoya as an example, students conceptually constructed the principles, contents, methods and other key elements of the guidelines for urban public space design, and further investigated Japanese urban design and spaces from five particular aspects: urban development in Nagoya, campus redevelopment, residential designs, heritage conservation and nature conservation with an attempt to understand the relationship between human environments and city from different perspectives.

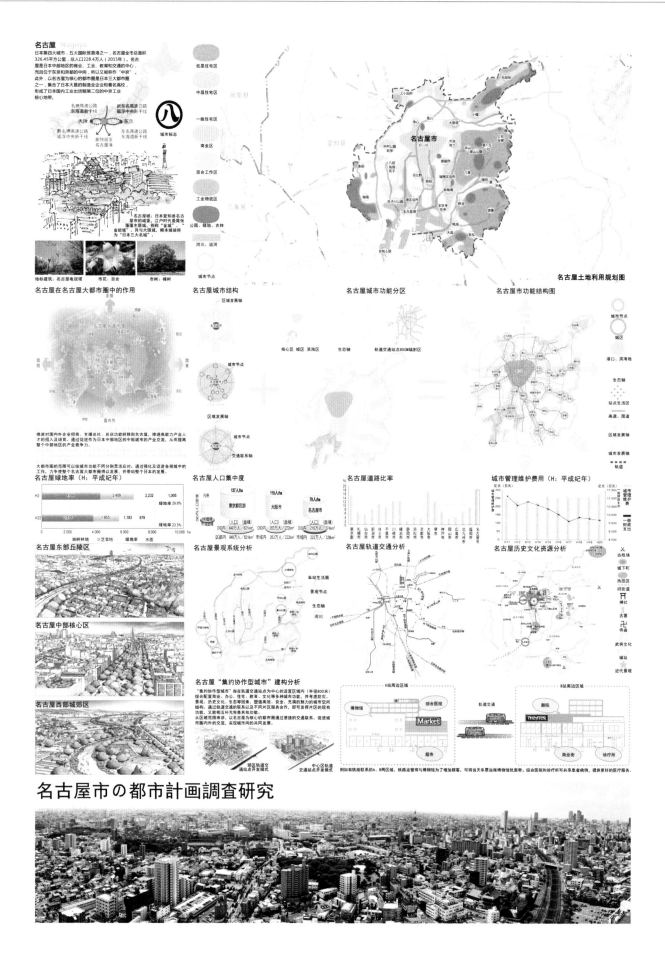

城市设计视角下的日本都市空间开发策略及实践调查研究

「中国と日本の建築」・古建筑

■ 奈良朱雀門与佛光寺大殿的比較研究

日本自古以来就有重视古代文化，文物并加以保护的传统，有大量的古代珍贵建筑遗物完整保存至今，成为日本乃至全世界的珍贵文化遗产。参观这些珍贵的古建筑，得到的第一个印象是保护得极好。这些国宝级的建筑都经过多次修理，很多还经过落架重修，但都严格遵守"整修如旧"的原则，在外观上基本看不到过分修补的痕迹，细意保持着历时千年的古貌。日本是很现代化的国家，但在这些国宝级建筑的周围，也尽可能创造出一种局部与世隔绝的环境，给观者提供一个领略千载前时代气息的短暂机会。在古建筑保护方面，日本有很多值得我们学习的地方。

奈良朱雀門

尽管日本在传入唐代风格的建筑后，和自己的传统文化和审美趣味相融合，不断有所创新，逐渐走上自己的发展道路，但在漫长的200多年中，唐代建筑也在发展变化，一次次随佛教传入日本的唐代建筑新风，也必然会在早期日本建筑中形成不同层次、不同程度的沉积和表现。

我国唐代建筑迄今只发现4座，数量远不能和日本的25座相比，4座中，最早的是山西五台南禅寺大殿，建于唐德宗建中三年（782年），约当日本奈良时代后期，和25座中最古的几座约略同时。实物，特别是较早实物的缺乏，给我们研究南北朝和唐前期建筑造成很大困难。但如果我们能对日本现存飞鸟、奈良时代遗构进行分析探索，发掘出在其中含蕴着的日本南北朝至唐代的遗风和规律，对于补充中国国内遗存实物不多的缺欠，充实我们对那个时代的建筑发展水平的认识，实在是有莫大的助益。

公元710年，日本都城迁到奈良平城京时，天皇的宫殿及政府机构就设在这里。794年，迁都到平安京（现在的京都）后，平城宫曾一度沦为农地，此后，随着兴福寺、东大寺等著名社寺和佛院的兴建，重新作为"寺院之都"而繁荣起来。1889年，奈良的宫殿遗迹首次被发现。目前，奈良在保持原有的广阔原野面貌的同时，根据发掘成果，复原了"朱雀门"（1989年复员）及庭园等，并继续进行着修复工作。

日本平城京即今天的奈良，是日本仿造唐长安建造的第一个都城。日本天皇将都城平城宫的正门命名为朱雀门，是古时外来使者、官员出入都城的入口。日本近代动用了极大的人力物力对其进行复原，花了11年的时间，直至2004年才修复完工。如今，朱雀门和大极殿一起，都是平城宫遗迹内主要的建筑。

朱雀门正立面　朱雀门侧立面　朱雀门柱头铺作　朱雀门补间铺作　朱雀门转角铺作

朱雀门为一层楼阁式歇山顶建筑，分六橑结构，面阔5间，进深8架椽，一层明间、次间明门，供通行。柱子具有明显卷杀，无普拍枋，与中国现存唐代建筑具有一致性。

朱雀门柱头铺作为七铺作斗拱，一、二跳偷心，最外跳施令栱。昂嘴斜度急促，与面阔方向平行。

朱雀门补间铺作以两层矮心柱及栱形成补间铺作。此法在日本较为普遍，招提寺金堂等均如此。国内现存唐宋木结构建筑无此做法。但辽代独乐寺山门山面补间铺作大斗下为一直真木，有相似之处。二层平坐为人字形补间铺作，两端为一斗三升。

佛光寺大殿

东大殿是佛光寺的正殿，在全寺最后的一重院落中，位置最高。此殿是由女弟子宁公遇施资、愿诚和尚主持佛光寺木构古佛光寺木构古门持、在原弥勒大阁的旧址上于唐大中十一年（公元857年）建成的。东大殿面阔七间、进深四间，用梁思成先生的话说，此殿"斗拱雄大，出檐深远"，是典型的唐代建筑。纵测量，斗拱断面尺寸为210X300厘米，是晚清斗拱断面的十倍；殿檐探出达三点九六米，这在宋以后的木结构建筑中也是找不到的。同时，大殿梁架的最上端用了三角形的人字架，这种梁架结构的使用时间，在全国现存的木结构建筑中列第一。

佛光寺大殿横架结构

佛光寺大殿正立面　佛光寺大殿侧立面　佛光寺大殿柱头铺作　佛光寺大殿补间铺作　佛光寺大殿转角铺作

佛光寺大殿为四阿顶殿堂建筑，金厢斗底槽结构，面阔7间，进深8架椽，柱子具有明显卷杀，无普拍枋。殿前面中央五间辟有板门，二尽端并开窗，其余三面以厚墙，仅山墙后部开有"扇面墙"。

佛光寺大殿柱头铺作为七杪双下昂五铺作，一、二、三跳偷心，最外跳施令栱，批竹昂。补间铺作为双秒双下昂五铺作，相比于朱雀门补间铺作的偷心造，变得复杂，有助于增强纵架稳定性。

朱雀门柱头斗栱相比于奈良时期的唐式建筑斗栱在形制上更加稳定、成熟。现存唐、宋、辽建筑均有很多相似性。

从分解图可以看出，奈良朱雀门和佛光寺大殿均为"层叠"式的殿堂构架形式。作为从唐代传入日本的早期木建筑，奈良朱雀门在建筑逻辑上与中国唐代木建筑具有一致性。

佛光寺大殿　奈良朱雀門

■ 奈良東大寺調查研究

东大寺南大门
东大寺山门
东大寺大殿

镰仓时期东大寺模型　七重塔模型

日本佛塔

东大寺（平假名とうだいじ）是日本华严宗大本山，又称大华严寺，金光明四天王护国寺。东大寺位于平城京（今奈良），是南都七大寺之一，距今约有一千二百余年的历史。

东大寺，1998年作为古奈良的历史遗迹的组成部分被列入世界文化遗产。佛寺是728年由信奉佛教的圣武天皇建立的。东大寺是全国68所国分寺的总寺院。因为建在首都平城京以东，所以被称作东大寺。

东大寺大佛殿，正面宽度57米，深50米，为世界最大的木造建筑。大佛殿中，放置着高15米以上的大佛像——卢舍那佛。

一个完整的日本寺庙要包含山门、金堂、讲堂、本坊、藏经阁、双塔等主要建筑元素。镰仓时期的东大寺，东西两侧还有两座七重塔，现在就只剩遗迹了。

■ 京都金閣寺調查研究

金阁寺

金阁寺（假名：きんかくじ），正式名称为鹿苑寺（假名：ろくおんじ），位于日本国京都府京都市北区，是一座临济宗相国寺派的寺院，为日本室町时代最具代表性的名园。金阁寺其名源自于日本室町时代著名的足利氏第三代幕府将军足利义满之法名，又因为内核心建筑"舍利殿"的外墙全是以金箔装饰，所以又称为"金阁寺"。今日我们所看到的舍利殿是昭和30年（1955年）时依照原样重新修复建造的，昭和62年（1987年）全殿外墙的金箔装饰皆全面换新，成为目前的状态。

1994年12月，鹿苑寺寺院被以"古都京都的文化财"的一部分被联合国教科文组织指定为世界文化遗产的重要历史建筑。

金阁寺（舍利殿）是一座紧邻镜湖池畔的三层楼阁状建筑，一楼是延续了当初藤原时代样貌的"法水院"（属寝殿风格，也是平安时代的贵族建筑风），二楼是镰仓时期的"潮音洞"（一种武家造，意指武士建筑风格）三楼则为中国（唐朝）风格的"究竟顶"（属禅宗佛殿建筑）。寺顶有宝塔状的造型，顶端栖息古样的金凤凰饰物。

三种不同时代不同的风格，却能在一栋建筑物上调和又完美，是金阁寺之所以受到瞩目的原因，除此之外，效仿自衣笠山的池泉回式庭园里有许多风格别致的日式造景，让金阁寺成为室町时代最具代表性的名园。

三岛由纪夫的长篇小说《金阁寺》取材于1950年金阁寺僧徒林养贤放火烧毁金阁寺的真实事件。振林养贤说他的犯罪动机是对金阁寺的美的嫉妒。《金阁寺》发表后大受好评，获第八届读卖文学奖。《金阁寺》的出版让更多人关注金阁寺，思考建筑遗产的保护问题。

城市设计视角下的日本都市空间开发策略及实践调查研究

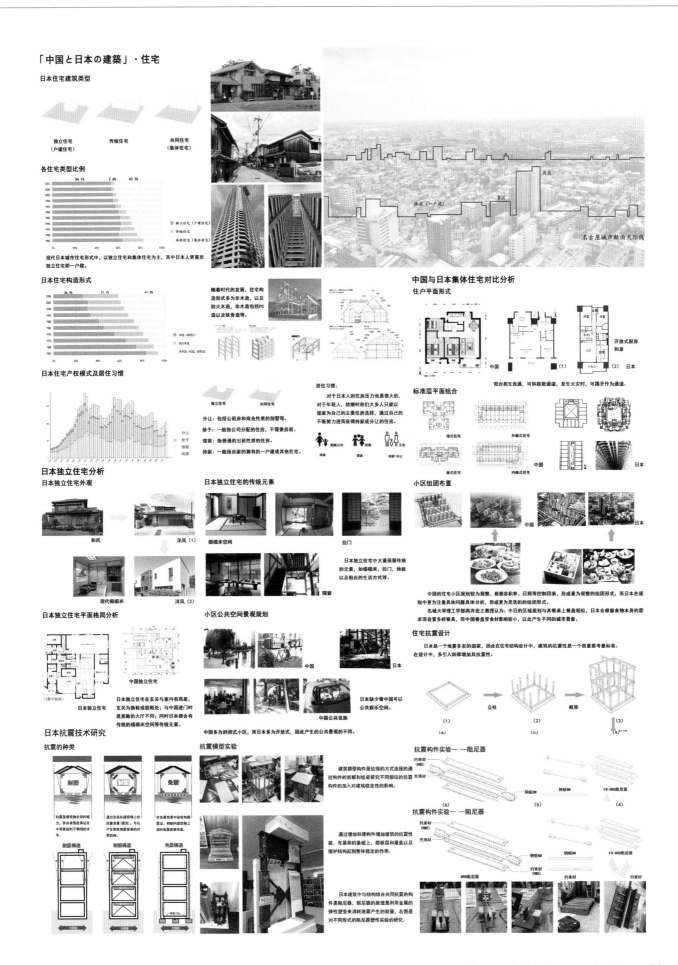

RESEARCH ON SUSTAINABLE DESIGN

04

可持续设计研究

新加坡国立大学联合工作营

Joint Workshop of the National University of Singapone

合作机构简介
Cooperative Institution Profile

新加坡国立大学，是新加坡首屈一指的世界级顶尖大学，为 AACSB 和 EQUIS 认证成员，亚洲大学联盟、亚太国际教育协会、国际研究型大学联盟、Universitas 21 大学联盟、环太平洋大学协会成员，在工程、生命科学及生物医学、社会科学及自然科学等领域的研究享有世界盛名。

National University of Singapore (NUS) is a leading world level university in Singapore with AACSB and EQUIS accreditation. As member of Asian Universities Alliance, Asia-Pacific Association for International Education, International Alliance of Research Universities, Universitas 21, and Association of Pacific Rim Universities, it has won a world reputation in a variety of research disciplines including engineering, life science and biomedicine, social sciences and natural sciences.

新加坡国立大学前身为 1905 年成立的海峡殖民地与马来亚联邦政府医学院；1912 年，该校改名为爱德华七世医科学校。1928 年，莱佛士学院成立。1949 年，爱德华七世医学院与莱佛士学院合并为马来亚大学。1955 年，新加坡华人社团组织创立了南洋大学。1962 年，马来亚大学位于新加坡的校区独立为新加坡大学。1980 年，新加坡大学和南洋大学合并，校名定为新加坡国立大学。

NUS developed out of the Straits Settlements and Federated Malay States Government Medical School which was founded in 1905. In 1912, the School was renamed King Edward VII Medical School. In 1928, the Raffles College was established. In 1949, Raffles College merged with King Edward VII College of Medicine to form the University of Malaya. In 1955, Chinese communities in Singapore founded Nanyang University. In 1962, the Singapore campus of the University of Malaya became an autonomous institution known as the University of Singapore. In 1980, the University of Singapore and Nanyang University merged to create the National University of Singapore.

根据 2015 年 4 月学校官网显示，学校建有肯特岗、武吉知马和欧南园 3 个校区；设有 16 所学院，包括一所音乐学院；有教学人员 2374 人，在校学生 37972 人，其中本科生 27975 人、研究生 9997 人。

According to official statistics from the University in April 2015, there are three campuses: Kent Ridge, Bukit Timah and Outram. The University has 16 Schools (17 by 2018, 13 Undergraduate and 4 Graduate), including one Music School; 2,374 teaching staff and 37,972 students, including 27,975 undergraduates and 9,997 graduates.

2017 年 10 月 25 日，2018 年全球最佳大学排行榜发布，新加坡国立大学在亚洲排名第一。2017 年 11 月，全球就业能力最强大学排行榜出炉，排名第 16。

In the QS World University Rankings 2018 published on Oct 25, 2017, NUS ranked first in Asia. In Nov 2017, NUS ranked 16th in the Graduate Employability Rankings 2018.

团队简介
Team Profile

指导教师
Instructors

刘少瑜
Stephen Lau

刘少瑜于 2015 年 1 月在新加坡国立大学设计与环境学院担任建筑系教授。
Stephen Lau joined the School of Design and Environment, National University Singapore in January 2015, as a Professor at the Department of Architecture.

刘少瑜曾在香港大学和巴特莱特建筑学院接受过建筑师培训，专攻可持续建筑及城市设计，具备建筑声学、灯光照明、绿色建筑评级和设计方面的专业知识和技能。他是香港注册建筑师、注册碳审核员，熟悉美国 LEED 评价标准、英国 BREEAM 评价标准，为 GBL 评审员（中国绿色建筑）。
Trained as an Architect (HKU and the Bartlett), he specializes in sustainable building and urban design. He has expertise in architectural acoustics, light and lighting, green building rating and design. He is a Registered Architect (HK), Certified Carbon Auditor, proficiency in LEED (U.S.A), BREEAM (U.K.), GBL Assessor (China Green Building).

他于 1985 至 2014 年任职于香港大学，担任院长、副院长（人力资源 / 资源处）、副院长（国际处）、执行主席、建筑学硕士协调人、博士生课程主任、副教授。在香港大学任职期间，他负责引领"适应性再利用与创意产业"和"可持续建筑与城市设计"建筑学硕士工作室，并指导了以"超密度城市现象学"为主题的建筑学硕士设计论文。以主要研究者身份完成了 60 个研究项目。主要研究领域为亚洲特大城市、以人为本的环境与建筑设计。指导近 25 个博士 / 硕士毕业，以及 9 个博士后研究员。在国际上，他是欧盟亚洲城市知识网络（UKNA）成员，参与了欧盟土地复合集约型利用研究项目（MILU）；是日本国际人居工程与设计学会（ISHED）会员，参与了 CIB 任务组 39- 特大城市项目；是加拿大国际建筑环境可持续发展计划（iiSBE）成员。在实践中，他作为国际 / 地方设计团队的合作者 / 负责人屡获奖项；于 2010-2014 年间担任大都会建筑事务所（OMA）亚洲区顾问和 RMJM 香港建筑设计事务所顾问，并参加了 2009 年的圣保罗双年展。
Previously at HKU (1985-2014) - Chair of Faculty, Associate Dean (HR/Resources) Associate Dean (International), Acting Head, M. Arch Coordinator, Director of the PhD Programme, Associate Professor. At HKU, he has conducted M. Arch studios on "adaptive reuse and the creative industry", "sustainable architecture and urban design"; and has supervised M. Arch. design theses on the themes of "the phenomenologies of the hyper-density city". He has completed sixty research projects as Principal Investigator. Main research: Asian Mega-cities, and human-oriented environmental and building design. Graduated nearly 25 PhD/MPhils and 9 Postdoc Fellows. Internationally, he was member to the European Union EU-Urban Knowledge Asia Network UKNA, the EU-Multiple and Intensive Land Use MILU Research Project, and the International Society of Habitat Engineering and Design ISHED (Japan), CIB Task Group 39- Megacities; International Initiatives on Sustainable Built Environment iiSBE (Canada). In practice, He has won awards, prizes, competitions as collaborators/leader of international/local design teams; advisor to OMA-AMO Asia (2010-14), and Robert Matthew Johnson-Marshall RMJM-HK. He exhibited at the Sao Paulo Biennale 2009.

曾任荷兰乌得勒支大学访问教授，新南威尔士大学高级访问研究员，中国多所大学的名誉教授。现任香港大学、同济大学、东南大学等高校的荣誉教授；同时为中国绿色建筑与节能委员会港澳分会副会长。
He has been Visiting Professor to the Utrecht University the Netherlands, Senior Visiting Fellow UNSW, and Honorary Professor to numerous universities in China. Currently, he is Honorary Professor to Hong Kong University, Tongji, Southeast, and others; he serves as Vice Chair of the China Green Building Council Hong Kong and Macau Chapters.

2015 年至今任职于新加坡国立大学，担任副校长、研究员、主任，负责博士课程、热带技术实验室、理学硕士综合可持续设计、建筑学高级研究中心。他负责 8 名硕博研究生，指导建筑学硕士论文，讲授气候响应型建筑、绿色建筑等课程，主持可持续发展研讨会等。在研究领域，他同时是两个光伏建筑一体化公共跨学科研究项目（近 475 万新元和 220 万新元）的主要研究者。此外，他还是城市噪声模型、自然定律和气候响应被动设计方面的其他研究项目（总计 30 万新元）和可再生能源、立面与自然通风这三个热带技术项目（约 40 万新元）的主要研究者和共同主要研究者。负责热带技术实验室（约 62 万新元）的设计和建设，并主管与新加坡两大房地产开发商的研究合作事务。目前，他的研究兴趣是以人为本的建筑与环境设计、建筑性能的物联网人机交互控制、建筑使用研究和生态城市主义。在专业领域，他是中国绿色建筑与节能专业委员会的活跃成员，并定期在中国举行演讲。
At NUS (2015-), he serves as Deputy Head- Research, Director: PhD Programme, Tropical Technologies Laboratory, M.Sc. Integrated Sustainable Design, Centre of Advanced Study of Architecture. He supervises research students (8 PhD/M.A. Research), M Arch theses, teaches modules in climatic responsive architecture, green buildings, and sustainability seminars. In research, he is Co-Principal Investigator of two public funded multidisciplinary research projects on Building Integrated Photovoltaics (approx. 4.75 Million and 2.2 Million SG dollars), he is also PI, Co PI of other research grants on urban noise modelling, biophilia and climatic responsive passive design, (totally SGD300,000) and three projects of the Tropical Technologies: renewable energies, productive façade and natural ventilation (around SGD 400,000). He is in charge of the design and construction of the Tropical Technologies Laboratory (around SGD 0.62 Million). He is in charge of research collaboration with two major real estate developers in Singapore. His current research interest is Human-oriented Building and Environment Design; Man-Machine Interactive Control (IoT) of Building Performance; Building Use Study; and Biophilic Urbanism. Professionally, he is an active member to the China Green Building Council and lectured regularly in China.

俞天琦
Yu Tianqi

北京建筑大学副教授，美国北卡罗来纳州立大学访问学者，哈尔滨工业大学博士。主要从事绿色建筑、复合建筑表皮等方向的理论及实践研究。主要承担绿色建筑设计、建筑学本科高年级及研究生的教学工作。主持科（教）研项目6项，主持并完成北京市自然科学基金项目；作为主要参与人，承担纵向科研10余项；受到北京市教委拔尖人才项目资助；指导学生参加国内外竞赛，获省部级以上奖励8项。主持参与大量建筑设计工程，已建成多项。出版学术专著《当代建筑表皮信息传播研究》，以第一作者发表学术论文20余篇。多次参编建筑设计资料集及行业规范。目前担任建筑学系副主任及党支部书记、"绿色建筑与节能技术"北京市重点实验室副主任。

Associate Professor at Beijing University of Civil Engineering and Architecture, visiting scholar at North Carolina State University, PhD at Harbin Institute of Technology. Her research focuses on the theory and practice of green architecture and composite building skins. She is in charge of teaching senior undergraduates and postgraduates majoring in green architecture design and architecture. She has undertaken 6 research (and teaching) projects, and completed several Beijing Natural Science Foundation projects. She is a principal contributor of over 10 government funded research projects, and has been awarded Top Talent Grant by Beijing Municipal Education Commission. She has instructed students in competitions both at home and abroad, and won 8 awards of and above provincial/ministry-level. She has been in charge of and contributed to numerous building projects, many of which have already been built. She has published monograph "Research on the Information Communication of Contemporary Architectural Skin", as well as over 20 academic papers as the first author. She has contributed to the editing of Architectural Design Instructions and professional codes. She is now deputy director and party branch secretary of Architectural Department, and deputy director of Beijing Key Lab for "Green Building and Energy Saving Technology".

郝石盟
Hao Shimeng

2010～2016年清华大学建筑学院建筑学建筑与技术研究所，获得工学博士；2013年荷兰代尔伏特工业大学，国家公派博士生联合培养；2008~2010年清华大学建筑学院城乡规划学，转提前攻读博士；2004~2008年清华大学建筑学院建筑学，获得建筑学学士。现任教于北京建筑大学。研究领域 可持续建筑与城市设计、建筑气候适应性、基于物理环境视角的民居研究。

Education background: From 2010 to 2016, she studied at Institute for Architecture and Technology, School of Architecture, Tsinghua University and obtained a PhD in Engineering; in 2013, she was member of a state-funded joint doctoral program at Delft University of Technology; from 2008 to 2010, she majored in urban and rural planning at School of Architecture, Tsinghua University and got an early doctoral degree; from 2004 to 2008, she learned architecture at School of Architecture, Tsinghua University and graduated as Bachelor of Architecture. She now teaches at Beijing University of Civil Engineering and Architecture. Research scope: sustainable building and urban design, building climactic adaptability, vernacular building studies based on physical environment.

袁超
Yuan Chao

袁超博士的研究兴趣集中在气候敏感地区的宜居和可持续城市规划和设计，特别是城市空气动力特性，这是高密度城市气候的关键组成部分，但往往难以模拟。他的研究目的是支持和发展现实生活中的实际规划和设计。袁博士曾参与香港的多项重大政策性研究项目，并积极参与中国（如武汉、澳门）和新加坡的一些项目。他开发了多个半经验模型，提供了连接建筑形态、景观、城市风环境和空气质量的理论知识。他的研究成果已广泛发表于各大科学期刊和书籍，并已纳入香港和武汉的城市气候研究，供城市规划者参考。

Dr. Yuan's research interests focus on climate sensitive urban planning and design for livable and sustainable cities, particularly on urban aerodynamic properties that are a critical part of high-density urban climate but often are difficult to simulate. His research effort is to support and develop the practical planning and design in the real life. Dr. Yuan has participated in several key policy-level research projects commissioned at Hong Kong, as well as actively involved in a few Chinese (e.g., Wuhan and Macau) and Singapore projects. He developed cluster of semi-empirical models, which provides an important knowledge linking the built morphology, landscape, and the city's wind environment and air quality. His research has been broadly published at leading scientific journals and books, and also incorporated into Hong Kong and Wuhan's urban climatic research for city planners' references.

刘兆杰
Lau Siu-kit Eddie

刘博士于1997年毕业于香港理工大学，获建筑服务工程学士学位，并于2003年获声学哲学博士学位。在攻读博士学位之前，他曾在建筑服务工程行业工作，担任香港陈炳祥工程顾问有限公司的顾问工程师。刘博士也曾在橡果国际有限公司、爱默生环境优化技术有限公司及德昌电机工业制造厂有限公司任职，积累了丰富的经验，为多个工业研究及发展项目提供专业技能，并为不同的系统设计了节能的解决方案。2004年至2009年，他在香港理工大学建筑服务工程学院担任访问学者和助理教授，之后在内布拉斯加大学林肯分校Charles W. Durham建筑工程与建设学院担任助理教授。他的多个研究项目得到了一致认可与支持。

Dr. Lau graduated from the Hong Kong Polytechnic University in 1997, with a Bachelor's degree in Building Services Engineering. He received his Doctor of Philosophy degree in Acoustics in 2003. Prior to taking up his doctoral program, Dr. Lau worked in the building services engineering industry as a consultant engineer at Daniel Chan & Associates Ltd. in Hong Kong. He also has had various experiences at Acron International Technology Ltd., Emerson Climate Technologies and Johnson Electric Industrial Manufactory, Ltd, where Dr Lau was able to contribute his expertise toward various industrial research and development projects, as well as devising solutions for various systems to be energy-efficient. From 2004 to 2009, he was a visiting lecturer and an assistant professor in the Department of Building Services Engineering at the Hong Kong Polytechnic University in Hong Kong, and went on to be an assistant professor at The Charles W. Durham School of Architectural Engineering and Construction of the University of Nebraska-Lincoln. His various research projects, have also garnered unanimous approval and support.

张冀
Zhang Ji

张冀是新加坡国立大学设计与环境学院亚洲城市可持续研究中心研究员。目前，他正在参与该中心的高密度阈值研究项目，通过数值模拟和评估，与研究组探讨城市形态、密度和环保成效之间的关系，以及城市模拟的理论、方法和技术。张冀拥有新加坡国立大学建筑学博士学位，方向为场所理论与环境心理学，并拥有华南理工大学建筑学硕士学位，方向为城市设计理论。他的研究兴趣包括：城市环境绩效模拟；城市开放空间规划与社区场所归属；建筑与城市设计。

Zhang Ji is a Research Fellow at the National University of Singapore's Centre for Sustainable Asian Cities (CSAC), within the School of Design and Environment (SDE). He is currently working on CSAC's High Density Threshold Studies research project in which he explores with the team the relationship between urban form, density and environmental performance through numerical simulation and evaluation, and the theory, methodology, and techniques for urban simulation. Zhang Ji holds a PhD in Architecture in the area of place theory and environmental psychology from the National University of Singapore and a Master of Architecture in the area of urban design theory from South China University of Technology. His research interests include urban environmental performance simulation; the relationship between urban open space planning and community place attachment; architecture and urban design.

团队学生
Team Students

岳梦迪

建筑学 – 研 2016 级

Yue Mengdi
Architecture - Postgrad 2016

高晶

建筑学 – 研 2016 级

Gao Jing
Architecture - Postgrad 2016

刘辉龙

建筑学 – 研 2016 级

Liu Huilong
Architecture - Postgrad 2016

张冠峰

建筑学 – 研 2016 级

Zhang Guanfeng
Architecture - Postgrad 2016

周培强

建筑学 – 研 2016 级

Zhou Peiqiang
Architecture - Postgrad 2016

李光耀

建筑学 – 研 2016 级

Li Guangyao
Architecture - Postgrad 2016

可持续设计研究

李天

建筑学 - 研 2016 级

Li Tian
Architecture - Postgrad 2016

吴茂梅

建筑学 - 研 2015 级

Wu Maomei
Architecture - Postgrad 2015

陈柏琳

建筑学 - 研 2015 级

Chen Bolin
Architecture - Postgrad 2015

TEAM

工作照
Teams at work

工作营任务书
Workshop Assignment

项目背景介绍
Project background introduction

屋顶是新加坡日常生活中热带建筑的重要元素，它不仅可以作为防止恶劣天气的住所，也是具有可见形式的特定文化的特征。典型例子是在马来西亚沿海地区的船屋。

A roof is an important element of a tropical building in the everyday life of Singapore. Not only does it act as a shelter against severe weather, it is also a signifier of the particular culture in a visible form. A typical example is the famous boat house in the Malayan coastal regions.

依据上述情况，学生们面临着对屋顶功能和表现的重新诠释的挑战，对肯特岭校园三个现有典型屋顶进行再设计。

The students are challenged to reinterpret the function and expression of a roof in the above contexts, with designs for the three existing typical roofs on the Kent Ridge campus.

设计任务及要求
Design tasks and requirements

工作营设计内容：新加坡国立大学肯特岗校区的三座屋顶。

The design Program:Three rooftops in Kent Ridge , Nus.

一处选址为AS8教学楼和中央图书馆的大型玻璃幕檐，该设计针对其所处的环境条件考虑不足，存在严重的使用问题，如眩光、辐射热和噪声反射等。

Site1: The massive glass canopy at the AS8 Teaching Building and the Central Library is criticized for its ignorance of the environmental function - resulting in serious practical problems, such as excessive and undesirable glare and heat, and reflecting noises.

第二处选址为设计与环境学院附近的食堂大屋顶，虽然其功能为遮阳、避雨，以及促进自然通风，但重新评估其作为捕风器或烟囱效应的有效性具有重要意义。

Site 2: For the extensive roof of the canteen at the Techno Edge Site near the School of Design and Environment, despite its intention to provide shield from heat gain, rain, and to allow for natural ventilation, it is pertinent to reevaluate its effectiveness as a wind catcher or purging hot air (stack effect).

第三处选址是设计与环境学院的一条风雨连廊，连接公共汽车站和设计与环境学院的正门。该连廊呈现给我们的问题是，其屋顶形式的选择是否与设计类学院的特质相契合？

Site 3: The third site is a roof canopy which links the bus stop to the main entrance of the School of Design and Environment. A question to be asked is whether the choice of the roof canopy is consistent to the character of a school of design?

项目信息
Project Information

新加坡气候特点：
Singapore Climate characteristics:

新加坡地处赤道，为热带雨林气候，常年气温变化不大，雨量充足，空气湿度高，日间相对湿度在60%~90%间，气候温暖潮湿，气温在25℃~37℃之间。4月、5月太阳辐射强而风速较弱，温度最高。
Singapore, located on the equator, is classified as tropical rainforest climate. Its climate is characterized by uniform temperature, high humidity and abundant rainfall. Daytime relative humidity ranges between 60% ~ 90%. The climate is warm and humid. The temperature hovers around a range of a minimum of 25°C and a maximum of 37°C. April is the hottest month of the year in Singapore, followed by May. This is due to light winds and strong sunshine.

新加坡有两个不同的季风季节，从12月到次年3月吹东北季候风，相当潮湿；6到9月吹西南季候风，较为干燥。两个季风期，间隔着季候风交替月，即4、5月和10、11月。在季候风交替月内，地面风弱多变，阳光酷热，形成下午至傍晚时分，全岛经常会有阵雨和雷雨。
Further contrasts are the monsoon seasons which happen twice each year. The first one is the humid Northeast Monsoon which occurs from November to next March. The second is the dry Southwest Monsoon season, from June to September. Periods between monsoon seasons receive less rain and wind, i.e. April and May, and October and November. In these months, there is mild and changing ground winds with scorching sunshine. Isolated to scattered showers occur in the late morning and early afternoon.

选址1：AS8楼和中央图书馆的大玻璃幕檐
Site 1: The massive glass canopy at the AS8 and the Central Library

该空间位于中央图书馆南侧，连接AS8教学楼与中央图书馆公交站，是校园内交通的重要节点。每日该空间会有大量的步行和车行流量经过，具有重要的交通等候与衔接功能。要求经过观察调研，重新设计该空间的顶部遮蔽形态。设计中要综合考虑到遮阳、避雨、采光等要求，最大程度的体现其空间特色与可持续性。
The space is located in the south of the central library, connecting the AS8 building and the central library bus stop, which is an important node for campus traffic. The space has a lot of pedestrian and vehicle flow every day, showing significant waiting and connecting needs. The task is to redesign the shape of the canopy for the space after observation and investigation, taking into account the shading, sheltering, lighting and other requirements, which reflects its spatial characteristics and sustainability on a maximum level.

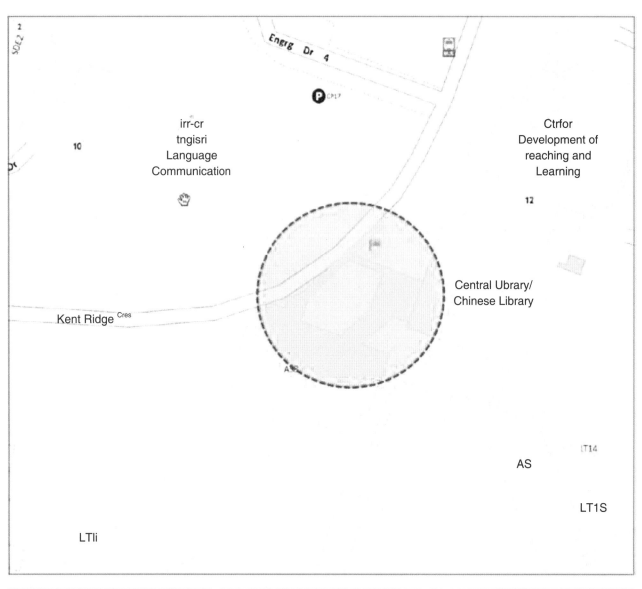

选址 2：工程学院餐厅

Site 2: The canteen at the Techno Edge Site

餐厅空间位于坡地，分成 3 个不同标高的地面，屋顶空间也由此形成高低错落的形式。该餐厅建成年代比较久，随着周边环境树木的生长，餐厅的周围环境也面临需要改善的问题。

The restaurant is located on a slope, having three different levels on the ground. The rooftop therefore has staggering heights. Since the restaurant was built a long time ago, with growth of trees in the surroundings, the environment of the restaurant demands improving.

要求对该就餐空间的屋顶进行重新设计，以更好地解决通风、采光、遮阳、避雨等问题，并结合周边环境，保留原有的生态系统的基础上，优化与周边环境的有机联系。

It is required that the roof of the dining space be redesigned to solve the problems of ventilation, lighting, sunshade, sheltering and other issues. It is also needed to, taking the surrounding environment into consideration, retain the original ecosystem and optimize the organic link with the environment.

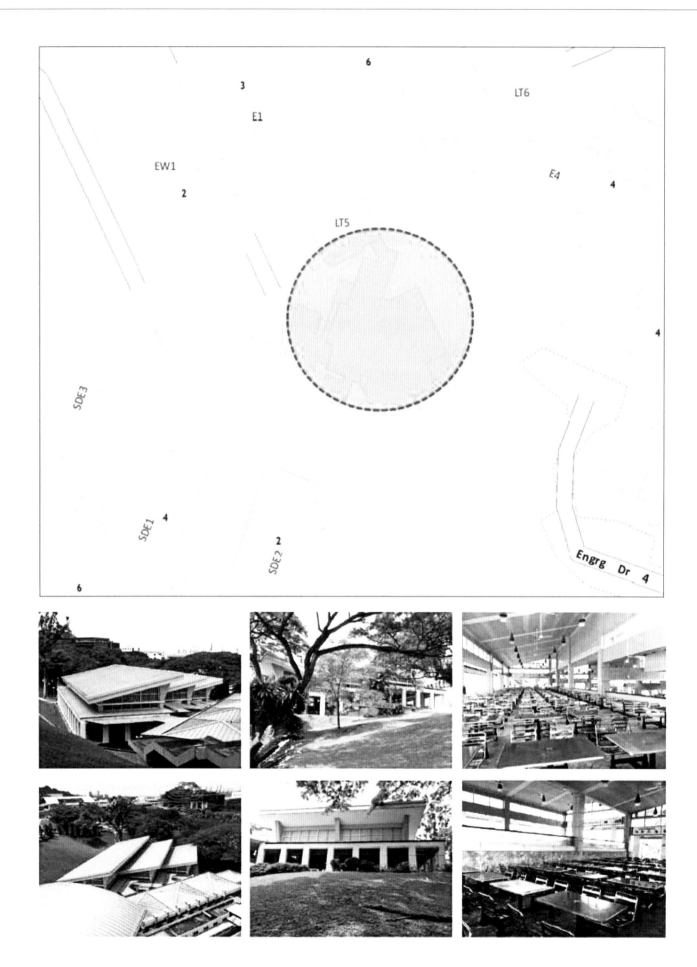

选址 3：设计与环境学院主入口与公交车站连廊空间
Site 3: Corridor space from the main entrance of SDE to bus stop

该空间是从校外空间进入学院的主要人行交通路线，巴士车站与学院大阶梯空间现由一个旧式连廊连接，地势由低到高，引导人流进入建筑内部。连廊中部，与一条机动车道交汇，行走过程中会有停留等待。通过对该处连廊的使用现状进行观察与分析，结合特殊的地形地势，对该空间进行重新设计。要求在基本交通功能的基础上，最大程度融合节能、生态、标志性及更多附加功能，赋予该空间更多活力和意义。

The space from the main steps of SDE to bus stop, as the principal pedestrian route from outside the campus to the School, is connected by an old-style corridor with a rising terrain that guides people to the inner space of the building. The middle of the corridor is intersected by a vehicle road, where people have to wait to cross. It is required to, based observation and analysis of the current situation of the corridor, redesign the space with regard to the special terrain. In addition to the basic traffic function, it is needed to maximize the integration of energy saving, ecology, iconic and more auxiliary features to fill the space with more vitality and meaning.

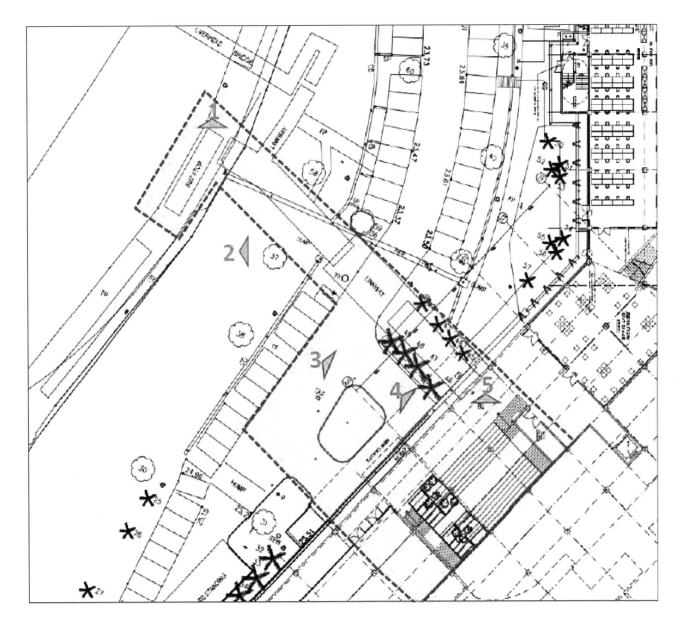

工作营组织方式
Workshop Organization

学生被分为三组，每组选择其中一处场地进行设计，要求最大地体现出对气候的响应，把绿色建筑与文化、社会、技术，以及建筑之间的关联与相互影响融合在设计之中。

The students are divided into teams of three, each working on one of the three sites, who will design accordingly to reflect foremost their appreciation of climate responsiveness, as well as the interplay between green architecture and culture, society, technology and architecture in their proposals.

提交文件为三份 A1 图纸，包括文本、插图、图表、轴测图等，要求清晰显示结构细节设计。此外，还需要通过计算机辅助设计手段，进行风流、热流、噪声、太阳路径等的模拟与分析。

The submission will include three A1 drawings - texts, illustrations, diagrams, and axonometric showing the constructional details. In addition, computer aided design measures should be taken, and wind and heat dynamics, noise, and sun path diagrams are required.

设计提交日期为 2017 年 8 月 19 日，展示介绍与公开评图，并颁发设计课程证书。

A certificate of attendance will be awarded on submission of the above documents by August 19, 2017, and successful presentation and review.

工作营日程
Workshop Schedule

时间 Time		任务 Task
第一天 Day 1	上午 Morning	交流活动：介绍 Communications: Introduction 现场参观 Site survey
	下午 Afternoon	环境评估教学，风、日光、噪声和太阳辐射得热 Education of environmental evaluation: wind, sunlight, noise and solar gain
第二天 Day 2	上午 Morning	计算机模拟辅助设计 Computer Aided Design simulations
	下午 Afternoon	计算机模拟辅助设计 Computer Aided Design simulations 三个屋顶设计（汇报基地调研，每组20分钟） Three rooftop designs (Presentation of site survey, 20min for each group) Techno Edge -参考书《House form and culture》 Techno Edge - reference《House form and culture》 AS8 - High Tech，参考墨菲西斯（Morphosis）建筑事务所的作品 AS8 - High Tech, reference works of Morphosis Architects SDE3雨棚—建筑系入口的表达 SDE3 Canopy – expression of entrance to Department of Architecture
第三天 Day 3	上午 Morning	专题设计 thematic design
	下午 Afternoon	
第四天 Day 4	上午 Morning	城市案例研究：参观Star Mall Urban case study: visit Star Mall
	下午 Afternoon	中期汇报 Mid-term presentation
第五天 Day 5	上午 Morning	设计成果绘制与制作 Design results drawing production
	下午 Afternoon	

续表

时间 Time		任务 Task
第六天 Day 6	上午 Morning	SDE汇报与评图 SDE presentation and review
	下午 Afternoon	
第七天 Day 7	上午 Morning	调研 Survey
	下午 Afternoon	
第八天 Day 8	上午 Morning	案例研究——新加坡热带建筑 Case study: tropical architecture in Singapore
	下午 Afternoon	
第九天 Day 9	上午 Morning	案例研究——新加坡热带建筑 Case study: tropical architecture in Singapore
	下午 Afternoon	

团队成果简介与展示
Team Achievements and Presentation

问题：设计地点为新加坡国立大学肯特岗校区环境与设计学院内的三个屋顶，分别为环境与设计学院入口雨棚、学生食堂和图书馆的入口雨棚。为了顺应学校的发展和体现学院的生态理念，需要对三处进行适应热带气候的改造和重建，并体现校园和地域文化。

Issue: The project site is the three rooftops of the School of Design and Environment at NUS Kent Ridge Campus – the canopy at the School's entrance, and the canopies at the students' Canteen and Library. In line with the development of the university and the ecological principle of the School, the three sites are to be adapted to tropical climate by transformation and reconstruction which reflect the campus and regional culture.

方法：针对周边环境中采光、通风和噪声的分析和改进是设计的重点，通过各项数据的调研和分析，以及对现状模型的数据进行软件计算得出的数据为基准，进行有针对性的设计处理，更好的实现数据指导下的绿色化设计。

Methodology: Analysis and improvement of environmental lighting, ventilation and noise is the key to design. Various data examinations and analyses, as well as data from software calculation of parametric models of current conditions, are used for specific design processing in order to better facilitate data-driven green design.

策略：设计方案以最基本的采光要求、自然通风降温需求和噪声控制为基准，通过流体计算软件Flow、声学设计软件Oden和grasshopper中honeybee与ladybug插件进行的光学模拟和气候模拟，来推进设计的进行。设计方案将针对某一项进行特色的优化设计，兼顾校园文化的同时，提供更具人性化的舒适性体验。

Strategy: The design is based on the fundamentals of lighting requirement, natural ventilation and cooling requirements and noise control. It proceeds with the fluid calculation software Flow, the acoustic design software Oden, and optic and climatic simulations with grasshopper plug-ins of honeybee and ladybug. The design focuses on one specific site for featured optimization, which respects the campus culture while providing more humane comfort.

设计：针对不同的地点，体现不同的文化和舒适性需求。食堂为学生较长时间聚集的地点，主要侧重自然通风降温的需求，提供更为舒适的就餐环境，两个设计分别突出了不同的风环境特色。学院入口雨棚则侧重于对学院特色和学生行为的参考，分别将光环境和声环境作为突出的重点，设计出创造性的特色学院入口空间。图书馆入口雨棚设计则更加侧重于交流与互动的公共空间氛围，完成了在环境优化的同时兼具活力的互动空间设计。

Design: Different culture and comfort needs are accounted for at different places. As a place where students gather for long time, the canteen is focused on natural ventilation and cooling requirements, which will create a more cozy dining environment. These two designs have dealt with different wind environment characteristics. The canopy at the School's entrance instead focuses on referencing the features of the School and students' behavior, prioritizing lighting and acoustic environment respectively, which has created a unique innovative School entrance space. The Library entrance canopy is designed to emphasize a communicative and interactive public space atmosphere, resulting in a design that is both an optimization of the environment and vigorous interactive space.

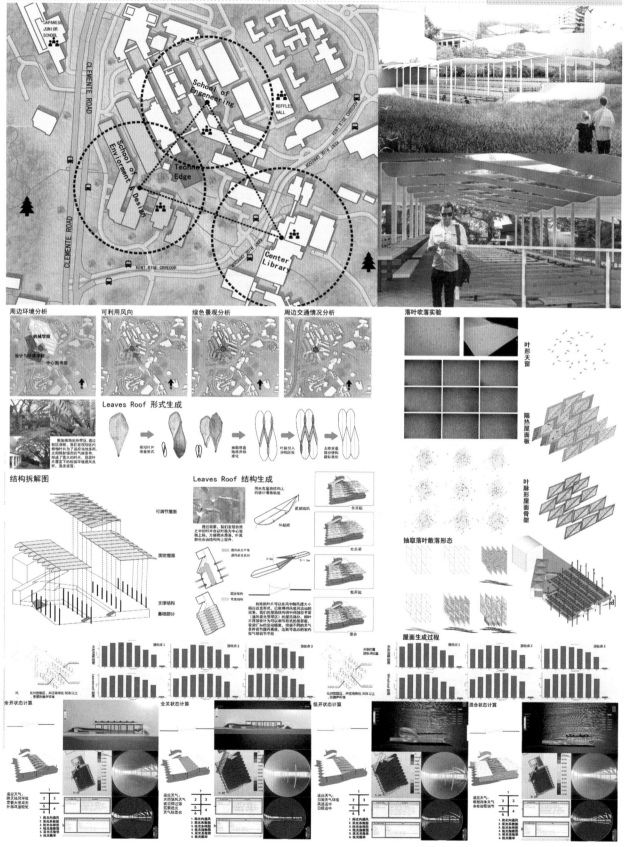

TECHNO EDGE 屋顶改造
——新加坡国立大学可持续设计工作坊

NUS SUSTAINABLE DESIGN WORKSHOP

LOCATION ANALYSIS

SINGAPORE_NUS　　TEDHND EDGE　　CANTEEN

SITE INVESTIGATION

此次进行研究的建筑位于新加坡国立大学建筑系区域，餐厅坐落半坡上，由三个依次抬高的就餐区以及一个取餐等候区组成，三个就餐区域的高差为1.5米。建筑屋顶样式为坡屋顶，经计算机模拟，即风环境模拟，光学模拟以及内部声源测试模拟。以及实地调研亲身体会用餐感受，发现现状建筑存在的一些问题：
1. 建筑形式方面，建筑本身的造型已已与周边环境格格不入，造型还是遵循上个世纪八十年代的形制
2. 结构方面，建筑体型并不轻盈，已不能体现建筑系本身的建筑性格
3. 功能方面，经过计算机声、光、风模拟以及实地就餐感受，发现餐厅较为吵闹，并且三层就餐区的标高增高，因此中间层餐厅的光之相对较弱，餐厅四周通透且建筑内部并未采用主动式空调系统，经过自然通风模拟分析，屋顶的改造会是基于被动式通风，优化通风设计
最后，随着时间的推移，建筑的形制以及功能无法满足人们的生活需求，本次由计算机模拟辅助设计，在现状建筑的基础上对其进行优化设计

在调研观察的过程中，也参观对比了新加坡其他的校区。学校建有肯特岗、武吉知马和欧南园3个校区，其中印象深刻的是U Town，University Town（简称UTown）是新加坡国立大学新建的一个学生生活区，于2011年秋季投入使用。采用综合性设计为学生提供了集学习与娱乐功能为一体的场所。

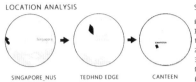

现状通风情况以及风速风流图

由风速计算机模拟flow design后，我们可以得出结果，新加坡当地风速约为7M/S，因此模拟到接近建筑表皮，可得出分数大约为4M/S，而根据数据我们可以得出，建筑中部就餐区域的通风情况不为适宜的舒适度，因此改造设计会一次为一个着重点。

现状餐厅内部噪声测试

噪声测试，是在餐厅三个不同标高的就餐地点选取声音接纳点，又根据餐厅实际就餐时产生的噪音而设置了发声点，测得的数据结果显示能量声压为53dB、57dB、58dB。显示就餐声环境不是特别舒适，因此设计着重此点。

新旧食堂之间的对比

肯特岗校区建筑系食堂与U town的异同点有很多，首先建筑本身都是采用被动式通风设计，这对用餐的舒适度有一个很大的提升，但是差异是肯特岗校区建筑在形制上稍微老旧一些。因此，这两个设计方案，不仅从分析风模拟以及噪声分析和炫光分析的等三个方面对建筑改造，在外在的建筑形制上也会与现代教学建筑相呼应。

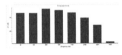

入口人视图

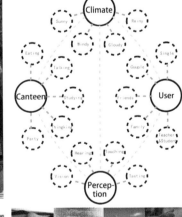

Climate — Sunny, Rainy, Windy, Cloudy
Canteen — Eating, Talking, Studying, Thinking, Party
User — Single, Double, Friends, Family, Teacher&Student
Perception — Hearing, Touching, Vision, Tasting

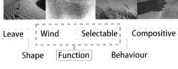

Leave　Wind　Selectable　Compositive
Shape　Function　Behaviour

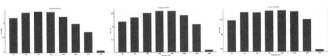
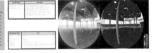

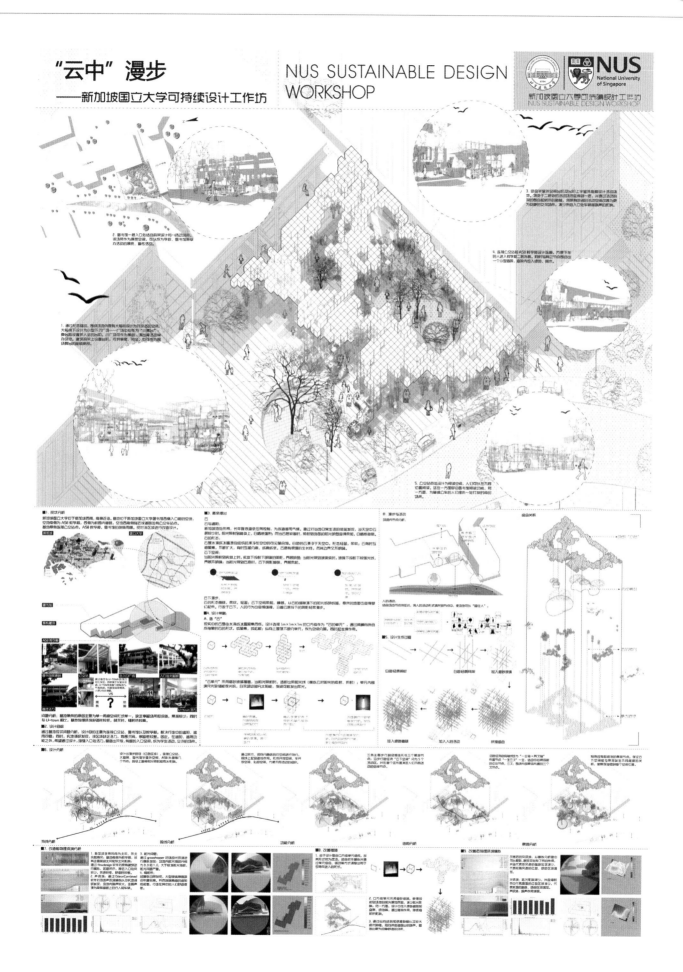

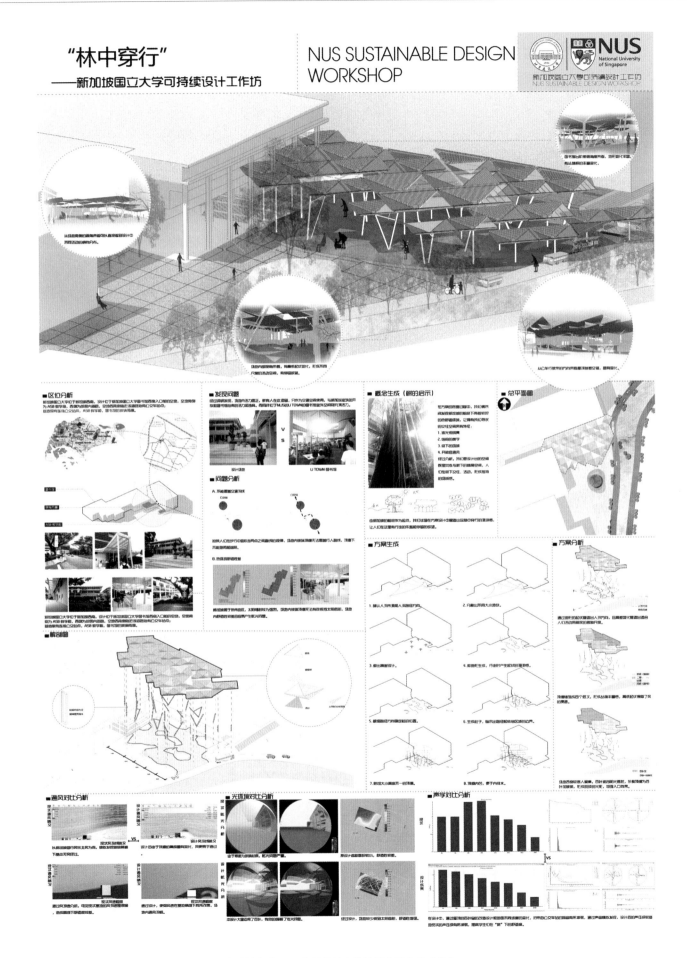

Shield of Vitality
——新加坡国立大学可持续设计工作坊

NUS SUSTAINABLE DESIGN WORKSHOP

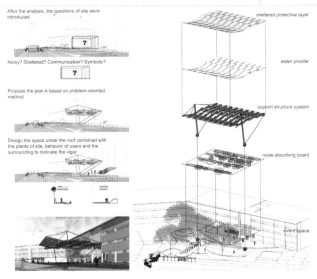

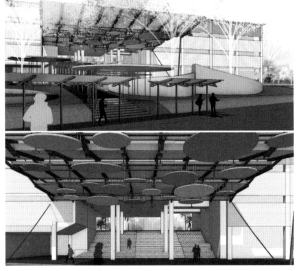

Airflow Analysis

The roof has a few influence on the air flow. Because of the autrium in the building, the airflow will go through much more easily.

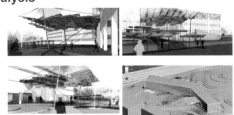

Acoustic Analysis

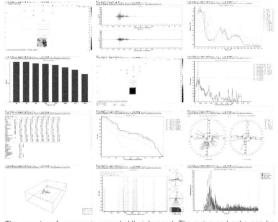

The sound performance is remarkablly inhenced. The noise and echo can be absorbed by the facilities design for sound.

Sunlight Analysis

The roof offer a shade space for students events.
The sunlight-radiation can be decrease in the tropical climate.

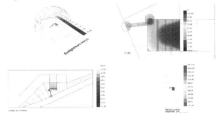

Design of SDE
新加坡国立大学可持续设计工作坊

NUS SUSTAINABLE DESIGN WORKSHOP

Site Analysis

The project is a conceptual design of corridor in the front of School of Design and Environment of National University of Singapore.
In tropical climate, the environment is the most important factor. We were trying to solve the problems by using the methods of performance-orientated design.

Airflow Analysis

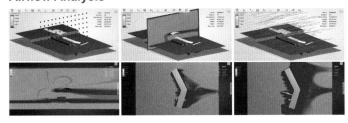

Acoustic Analysis

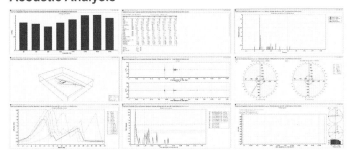

Sunlight Analysis

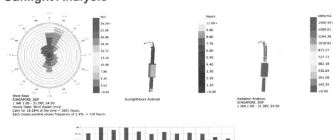

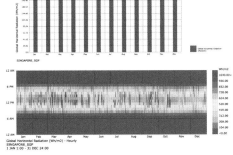

Human Flow & Forms Development

Radiant Umbrella

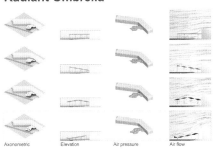

Axonometric   Elevation   Air pressure   Air flow

Shield of Vitality

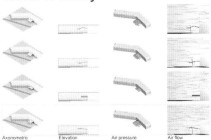

Axonometric   Elevation   Air pressure   Air flow

Performance-orientated Design

By analysing the current situation of the corridor, we found that there are few problems of the dragon-like roof canopy structure which links pedestrians from the Clement Road public bus stop to the main entrance of the SDE.

Problems:

Acoustic: Too much noise and echo
Sunlight: Unevenly distributed light and glare and the radiation make people feel uncomfortable
Airflow: The temperature of the roof is too high
Visual image: The characteristic of the corridor is too common and without any feature related to art.

Solutions:

Acoustic: Decrease noise and echo
Sunlight: Decrease the radiation which makes people feel uncomfortable
Airflow: Increase the ventilation
Visual image: Modern and elegant vision of art design

SOIL CONSTRUCTION PRACTICE

05

生土建构实践

"土生土长"国际生土联合工作营

"Back to Earth" - International Earthen Joint Workshop

合作机构简介
Cooperative Institution Profile

香港大学（HKU）是香港的一所公立研究型大学。成立于1911，其起源追溯到香港中医药学院，成立于1887。它是香港最古老的高等学府。是一所位处中华人民共和国香港特别行政区的国际化公立研究型大学。

Hong Kong University is a public research university in Hong Kong. Established in 1911, it could be traced back to Hong Kong College of Traditional Chinese Medicine which was founded in 1887. It is the most time-honored institution of higher learning in Hong Kong, and an international public research university in Hong Kong Special Administrative Region of the People's Republic of China.

香港大学被普遍认为是世界上最国际化的大学之一，同时也被认为是亚太地区最负盛名的大学之一，有亚洲"常春藤"之称。

Crowned as Asia's "Ivy League" university, it is widely regarded as one of the most international universities in the world and is among the most prestigious universities in the Asia-Pacific region.

香港大学有十所学院，以英语为教学语言。它在经济、金融、会计、生物医学、牙医、教育学、人文学科、法学、语言学、政治学与社会科学等领域展现出较强的科研实力。香港大学也是世界上第一个成功分离SARS病原体冠状病毒的研究小组。

Hong Kong University has ten schools, with English being its teaching language. It has demonstrated strong scientific research strength in such fields as economics, finance, accounting, biomedicine, dentistry, education, humanities, law, linguistics, political science and social science. It also has the world's first research team that has successfully isolated SARS pathogen coronavirus.

无止桥慈善基金于2007年在香港成立，旨在鼓励香港和内地大学生运用环保理念，义务为国内贫困和偏远的农村设计和修建便桥及村庄设施，并建立两地学生和村民之间的「心桥」。借以促进香港和内地的沟通、了解与融和；改善内地偏远、贫困农村的生活环境和质素；启发社会尊重、欣赏和保护地方文化、传统和环境，提倡可持续理念；提供大学生服务学习的机会，亲身为村民带来实质改变。

Founded in Hong Kong in 2007, Wu Zhi Qiao (Bridge to China) Charitable Foundation aims to encourage college students in Hong Kong and the Chinese mainland to apply the concept of environmental protection to design and build auxiliary bridges and village facilities for poor and remote rural areas of Chinese mainland, and to build a "heart bridge" between students and villagers of the two regions. It strives to facilitate communication, understanding and integration between Hong Kong and the Chinese mainland; to improve the living environment and quality of remote and poor rural areas in the Chinese mainland; to enlighten the society to respect, appreciate and protect local culture, traditions and environment, and advocate sustainability concept; and to provide college students with opportunities to serve and learn and bring about material changes to villagers.

基金至今已发动超过3,200名大学生和专业人士，在内地偏远农村地区完成近60个建设项目，包括44座无止桥、3个村民活动中心、3个农村示范项目和一系列民生改善项目，受惠人数超过60,000人次。部分项目更获得专业机构和国际认同，当中包括环保建筑大奖和联合国教科文组织亚太区文物古迹保护奖。

The Foundation has so far mobilized more than 3200 college students and professionals to complete nearly 60 construction programs in the remote rural areas of the Chinese mainland, including 44 Wu Zhi Qiao programs, 3 villager activity centers, 3 rural demonstration programs and a series of programs to improve people's living, which have benefited more than 60,000 people. Some of the programs have been recognized by specialized institutions and the world, including the Green Building Award and UNESCO Asia-Pacific Heritage Awards for Culture Heritage Conservation.

团队简介
Team Profile

指导教师
Instructors

穆钧
Mu Jun

香港中文大学博士，曾任西安建筑科技大学建筑学系主任，现任北京建筑大学教授、博导，兼任住建部村镇司生土建筑研究与发展中心常务副主任、住建部传统民居保护专家委员会副主任委员、联合国教科文组织"生土建筑、文化与可持续发展教席"负责人、中国建筑学会生土建筑分会常务理事兼副秘书长、无止桥慈善基金项目委员会主席（内地）等职务。

PhD of Chinese University of Hong Kong, has ever been the dean of School of Architecture of Xi'an University of Architecture and Technology, and now is the professor and doctoral supervisor of Beijing University of Civil Engineering and Architecture, and executive deputy director of Earthen Architecture Research and Development Center of Department of Village and Township Construction of Ministry of Housing and Urban-Rural Development of the People's Republic of China (MOHURD), vice director of Expert Committee on the Protection of Traditional Residences of MOHURD, the responsible person of the UNESCO Chair "Earthen architectures, constructive cultures and sustainable development", standing director and deputy secretary general of the Earthen Architectures Branch of the Architectural Society of China, and chairman of Wu Zhi Qiao Charitable Foundation program committee (Chinese mainland), etc.

蒋蔚
Jiang Wei

北京建筑大学讲师，土上建筑主持建筑师。法国国家注册建筑师，法国巴黎拉维莱特国立建筑学院硕士，曾任教于西安建筑科技大学，现任教于北京建筑大学。

Personal profile: Jiang Wei, lecturer of Beijing University of Civil Engineering and Architecture, and presiding architect of earthen architectures. He is the registered architect in France, master of ENSAPLV. He ever taught at Xi'an University of Architecture and Technology, and now teaches at Beijing University of Civil Engineering and Architecture.

任中琦
Ren Zhongqi

荷兰贝尔拉格建筑学院硕士，曾就职于荷兰OMA建筑事务所，曾任教于香港中文大学和西安建筑科技大学，现任教于北京建筑大学。

Personal profile: Ren Zhongqi, master of the Berlage Institute. He ever worked at Office for Metropolitan Architecture (OMA) in Netherlands and taught at Chinese University of Hong Kong and Xi'an University of Architecture and Technology. Now he teaches at Beijing University of Civil Engineering and Architecture.

郝石盟
Hao Shimeng

2010~2016年清华大学建筑学院建筑学建筑与技术研究所，获得工学博士；2013年荷兰代尔伏特工业大学，国家公派博士生联合培养；2008~2010年清华大学建筑学院城乡规划学，转提前攻读博士；2004~2008年清华大学建筑学院建筑学，获得建筑学学士。现任教于北京建筑大学。

Personal profile: Hao Shimeng. From 2010 to 2016, she studied at Institute for Architecture and Technology, School of Architecture, Tsinghua University and obtained a PhD in Engineering; in 2013, she was member of a state-funded joint doctoral program at Delft University of Technology; from 2008 to 2010, she majored in urban and rural planning at School of Architecture, Tsinghua University and got an early doctoral degree; from 2004 to 2008, she learned architecture at School of Architecture, Tsinghua University and graduated as Bachelor of Architecture. She now teaches at Beijing University of Civil Engineering and Architecture.

团队学生
Team students

Alexandra Viscusi

普林斯顿大学 - 建筑学 - 本 2016 级

Alexandra Viscusi
Princeton University - Architecture -
Undergrad 2016

Lindsey Swartz

普林斯顿大学 - 土木工程 - 本 2016 级

Lindsey Swartz
Princeton University - Civil engineering -
Undergrad 2016

Teresa Irigoyen-Lopez

普林斯顿大学 - 建筑学 - 本 2014 级

Teresa Irigoyen-Lopez
Princeton University - Architecture -
Undergrad 2014

郑启诺

纽卡斯特大学 - 建筑学 - 本 2014 级

Zheng Qinuo
Newcastle University - Architecture -
Undergrad 2014

吴曦婷

香港理工大学 - 建筑学 - 本 2015 级

Wu Xiting
Hong Kong Polytechnic University -
Architecture - Undergrad 2015

林泽昕

香港理工大学 - 建筑学 - 本 2015 级

Lin Zexin
Hong Kong Polytechnic University -
Architecture - Undergrad 2015

陈盈盈

香港理工大学 - 工程项目管理 - 本 2015 级

Chen Yingying
Hong Kong Polytechnic University - Engineering
Project Management - Undergrad 2015

张威宁

北京建筑大学 - 供热、供燃气、通风及空调
工程 - 研 2016 级

Zhang Weining
Beijing University of Civil Engineering and
Architecture - Heating, Gas Supply, Ventilating
and Air Conditioning - Postgrad 2016

朱轩宇

北京建筑大学 - 建筑学 - 本 2015 级

Zhu Xuanyu
Beijing University of Civil Engineering and
Architecture - Architecture - Undergrad 2015

杨洋

北京建筑大学 - 建筑学 - 研 2017 级

Yang Yang
Beijing University of Civil Engineering and Architecture - Architecture - Postgrad 2017

李广林

北京建筑大学 - 建筑学 - 研 2017 级

Li Guanglin
Beijing University of Civil Engineering and Architecture - Architecture - Postgrad 2017

岳海波

中南大学 - 建筑学 - 本 2015 级

Yue Haibo
Zhongnan University - Architecture - Undergrad 2015

刘雨橦

吉林建筑大学 - 景观园林 - 本 2015 级

Liu Yutong
Jilin Jianzhu University - Landscape Architecture - Undergrad 2015

顾倩倩

西安建筑科技大学 - 建筑学 - 研 2017 级

Gu Qianqian
Xi'an University of Architecture and Technology - Architecture - Postgrad 2017

周全

西安建筑科技大学 - 建筑学 - 本 2012 级

Zhou Quan
Xi'an University of Architecture and Technology - Architecture - Undergrad 2012

史雨佳

西安建筑科技大学 - 建筑学 - 研 2017 级

Shi Yujia
Xi'an University of Architecture and Technology - Architecture - Postgrad 2017

姚雅露

西安建筑科技大学 - 建筑学 - 研 2017 级

Yao Yalu
Xi'an University of Architecture and Technology - Architecture - Postgrad 2017

高志鹏

西安建筑科技大学 - 建筑学 - 研 2017 级

Gao Zhipeng
Xi'an University of Architecture and Technology - Architecture - Postgrad 2017

杨亚杰

西安建筑科技大学 - 建筑学 - 研 2017 级

Yang Yajie
Xi'an University of Architecture and Technology - Architecture - Postgrad 2017

谢月皎

西安建筑科技大学 - 建筑学 - 研 2016 级

Xie Yuejiao
Xi'an University of Architecture and Technology - Architecture - Postgrad 2016

韩霈雯

西安建筑科技大学 - 建筑学 - 本 2016 级

Han Peiwen
Xi'an University of Architecture and Technology - Architecture - Undergrad 2016

刘寒露

西安建筑科技大学 - 建筑学 - 本 2016 级

Liu Hanlu
Xi'an University of Architecture and Technology - Architecture - Undergrad 2016

马列

西安建筑科技大学 - 建筑学 - 本 2016 级

Ma Lie
Xi'an University of Architecture and Technology - Architecture - Undergrad 2016

杨茹

西安建筑科技大学 - 建筑学 - 研 2016 级

Yang Ru
Xi'an University of Architecture and Technology - Architecture - Postgrad 2016

周健

西安建筑科技大学 - 建筑学 - 研 2016 级

Zhou Jian
Xi'an University of Architecture and Technology - Architecture - Postgrad 2016

胡亮

西安建筑科技大学 - 建筑学 - 研 2016 级

Hu Liang
Xi'an University of Architecture and Technology - Architecture - Postgrad 2016

郭雷平

西安建筑科技大学 - 建筑学 - 研 2016 级

Guo Leiping
Xi'an University of Architecture and Technology - Architecture - Postgrad 2016

王森华

西安建筑科技大学 – 建筑学 – 研 2015 级

Wang Senhua
Xi'an University of Architecture and Technology - Architecture - Postgrad 2015

王博航

西安建筑科技大学 – 建筑学 – 研 2015 级

Wang Bohang
Xi'an University of Architecture and Technology - Architecture - Postgrad 2015

赵宜芊

西安建筑科技大学 – 建筑学 – 研 2015 级

Zhao Yiqian
Xi'an University of Architecture and Technology - Architecture - Postgrad 2015

杨琪

西安建筑科技大学 – 建筑学 – 研 2015 级

Yang Qi
Xi'an University of Architecture and Technology - Architecture - Postgrad 2015

邓博伟

西安建筑科技大学 – 建筑学 – 研 2015 级

Deng Bowei
Xi'an University of Architecture and Technology - Architecture - Postgrad 2015

赵西子

西安建筑科技大学 – 建筑学 – 研 2015 级

Zhao Xizi
Xi'an University of Architecture and Technology - Architecture - Postgrad 2015

师仲霖

西安建筑科技大学 – 建筑学 – 研 2016 级

Shi Zhonglin
Xi'an University of Architecture and Technology - Architecture - Postgrad 2016

詹林鑫

西安建筑科技大学 – 建筑学 – 研 2014 级

Zhan Linxin
Xi'an University of Architecture and Technology - Architecture - Postgrad 2014

张浩

西安建筑科技大学 – 建筑学 – 研 2013 级

Zhang Hao
Xi'an University of Architecture and Technology - Architecture - Postgrad 2013

团队合影
Team photos

工作照
Teams at work

工作营任务书
Workshop Assignment

项目背景介绍
Project background introduction

在本次"土生土长"工作营中，志愿者可以充分利用不同原色的生土为材，以大量的小型试件的形式，分别在强度、色彩、肌理、形态等层面，进行开放式的设计与制作尝试。

In the "Back to Earth" workshop, volunteers will be invited to make full use of earth of different primary colors, design and make all kinds of products in small samples with different shades of strength, colors, textures, and shapes by using their imagination freely.

志愿者们可以观察、触摸、思考以及想象，以此来感受一种"新"的传统材料，及其潜在的设计语言、表现形式与相适宜的应用定位。另外，志愿者们将会深入参与生土建筑实践京港双城展的前期准备，参与生土展品的设计与制作、展板设计及排版、协助布展等工作，在自身专业的基础上，通过动手实际操作和深度交流，拓展和深化志愿者在绿色建筑、乡土营建、生土建筑等方面的专业认知和视野。

Volunteers will feel a "new" traditional material and its potential design language, expression form and suitable application orientation by observation, touching, thinking and imagination. In addition, they will be involved in the preparation for the Earthen Architecture Beijing-Hong Kong Bi-City Exhibition, participate in the designing and making of earthen exhibits and the styling and composing of display boards, and provide assistance for exhibition arrangement. Based on their professional knowledge, they will expand and deepen their recognition and vision on green buildings, rural construction and earthen architectures through practical operation and in-depth communication.

以土为材，是我国乃至全世界历史最为悠久、应用最为广泛的营造传统之一，但传统生土民居存在一些相对缺陷，难以满足当下人们的需求。过去十多年间，在住房和城乡建设部村镇司与无止桥慈善基金的大力推动和支持下，联合国教科文组织"生土建筑、文化与可持续发展"教席国际研究网络的支援下，穆钧教授研究团队针对传统生土营建工艺的发掘、改良与革新，取得了一系列广受关注和令人鼓舞的成果。本次工作营所筹办的展览旨在使人们重新认识我国生土营建的传统，在了解现代生土材料科学的同时，思考与审视以生土为代表的传统营建工艺在今天的应用潜力与相适宜的发展定位。

"Taking earth as material" is one of the most time-honored and popular tradition in China and the world. However, there are some defects in traditional earthen architectures which make them hard to meet people's needs. In the last decade, with the advancement and support of the Department of Village and Township Construction of Ministry of Housing and Urban-Rural Development of the People's Republic of China (MOHURD) and Wu Zhi Qiao (Bridge to China) Charitable Foundation, as well as the UNESCO Chair "Earthen architectures, constructive cultures and sustainable development" international research network, the research team led by Prof. Mu Jun has achieved a series of widely recognized and inspiring results in the exploration, improvement and innovation of traditional earthen architecture technique. The exhibition organized by the workshop aims to help people re-recognize the tradition of earthen construction of China, and think and examine the present application potential and suitable development orientation of traditional construction techniques represented by earth while get to know modern earthen material science.

设计任务及要求
Design tasks and requirements

于 2017 年 8 月底完成展板设计、展品制作、展厅布置。展览于 2017 年 9 月 16 日在北京建筑大学西城校区创空间展厅正式开幕，在北京建筑大学西城校区大学生活动中心开展现代生土建筑国际论坛。

The design of the display boards, the production of exhibits and the arrangement of the exhibition hall shall be completed by the end of August. The exhibition will be opened officially in Creation Space, Xicheng Campus, Beijing University of Civil Engineering and Architecture on September 16, 2017. The International Modern Earthen Architecture Symposium will be held at the Student Activity Center of Xicheng Campus of the university.

工作营组织单位
Workshop organization

工作营主办单位：住房和城乡建设部村镇司、无止桥慈善基金
Workshop hosts: Department of Village and Township Construction of Ministry of Housing and Urban-Rural Development of the People's Republic of China (MOHURD), Wu Zhi Qiao (Bridge to China) Charitable Foundation

工作营承办单位：北京建筑大学建筑与城市规划学院、北京未来城市设计高精尖创新中心
Workshop organizers: School of Architecture and Urban Planning, Beijing Advanced Innovation Center for Future Urban Design of Beijing University of Civil Engineering and Architecture

工作营地点：北京建筑大学创空间展厅、教五教室
Workshop location: Creation Space Exhibition Hall and No. 5 Teaching Building of Beijing University of Civil Engineering and Architecture

团队成果简介与展示
Team Achievements and Presentation

本次工作营是由无止桥慈善基金与北京建筑大学于 2017 年夏季联合举办的，是为期 2 个月的大学生暑期工作营。工作营的内容是筹办生土展览，所有的实物展品由来自内地、香港和美国的 30 余位大学生志愿者和实习生，在研究团队的带领下共同协作，亲手完成。展览于 2017 年 9 月开幕，正值无止桥慈善基金成立十周年、香港回归二十周年之际，在住房和城乡建设部村镇司、无止桥慈善基金、北京建筑大学的支持下，团队于北京建筑大学举办了我国首次以生土建筑为主题的专题展览。本次展览在总结过往十年的同时，联合国际现代生土建筑界的同仁与先行者，通过图文展板、实物工具、样品试件、片段装置、建筑模型、视频资料等形式，首次较为全面地呈现了中国传统生土民居建筑及其建造技术、生土材料应用基本科学原理、生土材料美学表现、团队在现代生土建筑领域的实践与探索，以及国际当代生土建筑优秀案例等板块的内容，旨在使人们重新认识我国生土营建的传统，在了解现代生土材料科学的同时，思考与审视以生土为代表的传统营建工艺在今天的应用潜力与相适宜的发展定位。展览于 2017 年 9 月 16 日正式开幕，11 月 16 日闭幕。尽管展览受空间、时间与团队经验的限制，难以充分系统地呈现生土建筑领域的方方面面，但有幸迎来了来自全国各地的 6000 余位参观者，也有幸获得了建筑界乃至社会各界的广泛关注和肯定。这次展览向社会大众呈现一幅以"生土"绘就的画卷，抛砖引玉，开启一扇回望传统、审视传承的窗口。

Co-hosted by Wu Zhi Qiao Charitable Foundation and Beijing University of Civil Engineering and Architecture in the summer of 2017, this summer vacation workshop lasts for 2 months. Its work is to prepare for the earth exhibition. All exhibits are finished together by more than 30 college volunteers and interns from the Chinese mainland, Hong Kong and the US guided by the research team. On the occasion of the 10th anniversary of the establishment of Wu Zhi Qiao Charitable Foundation and the 20th anniversary of the return of Hong Kong to the motherland, the exhibition was opened in September 2017. With the support of the Department of Village and Township Construction of MOHURD, Wu Zhi Qiao Charitable Foundation and Beijing University of Civil Engineering and Architecture, the team held China's first special exhibition themed with earthen architecture at Beijing University of Civil Engineering and Architecture. Earthen architectures in the past decade was reviewed at the exhibition. With the help of fellows and pioneers of the field of international modern earthen architectures, the exhibition comprehensively presented Chinese traditional earthen architectures and its construction techniques, basic scientific principles of the application of earthen material, aesthetic expression of earthen material, the team's practice and exploration in the field of modern earthen architectures, and the excellent cases of international modern earthen architectures for the first time in the forms of graphic boards, physical tools, samples, fragment devices, building models, video data, etc. It helps people learn Chinese tradition of earthen construction again and think and examine the present application potential and suitable development orientation of traditional construction techniques represented by earth while get to know modern earthen material science. The exhibition was opened officially on September 16 and closed on November 16, 2017. Although the exhibition is limited by space, time and team experience, and it is difficult to fully and systematically present all aspects of the field of earthen architectures, it is fortunate that the exhibition had ushered in more than 6,000 visitors from all over the country, and won wide attention and recognition from the field of architecture and even all sectors of society. This exhibition has presented to the public a panorama of "earth", which opens a window for people to look back on the tradition and examine inheritance.

生土建构实践

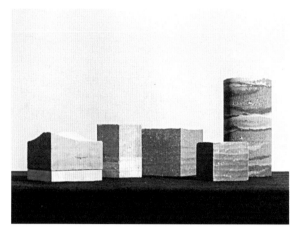

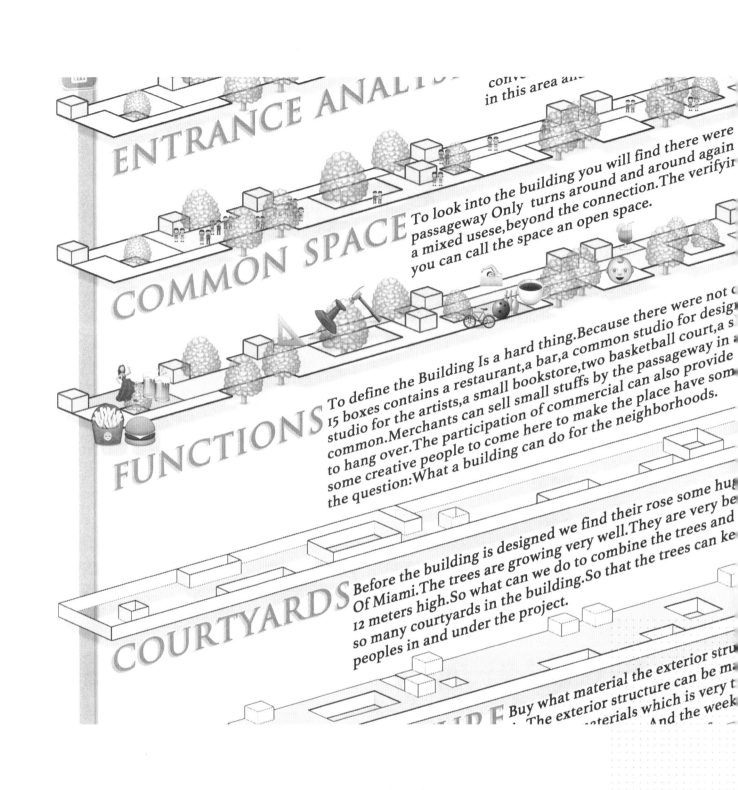

TROPICAL CITIES AND ARCHITECTURE: URBAN DESIGN FOR FLEXIBLE COASTAL LINES

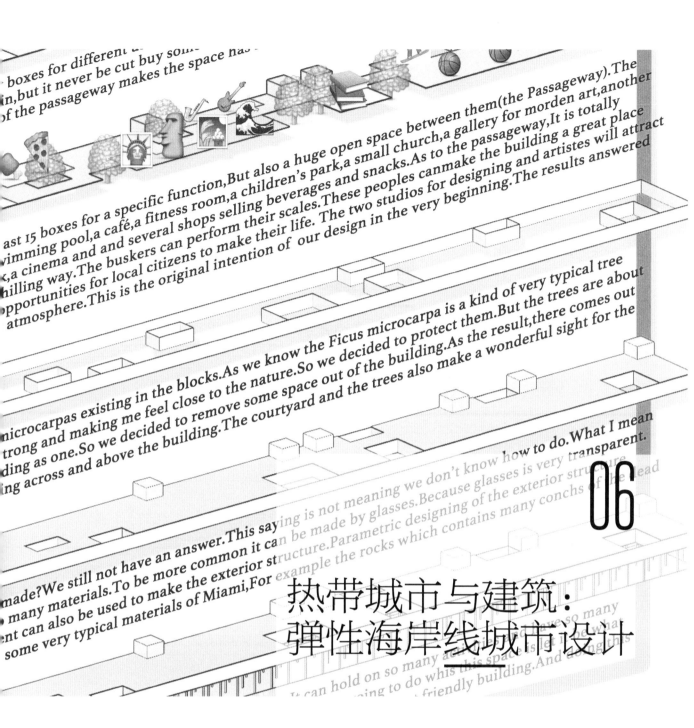

热带城市与建筑：
弹性海岸线城市设计

美国迈阿密大学联合工作营

Joint Workshop of the University of Miami

合作机构简介
Cooperative Institution Profile

美国迈阿密大学,简称UM,位于美国佛罗里达州的南部小城科勒尔盖布尔斯,是一所始建于1925年的美国顶尖私立综合大学,被公认为佛罗里达州最负盛名的大学。2017年,迈阿密大学在U.S.News美国大学综合排名中位列全美第44名。

University of Miami, Abbreviated as UM and located in Coral Gables, a small city in southern Florida, University of Miami is a top private comprehensive university in the US. Founded in 1925, UM is generally recognized as the most renowned university in Florida. UM ranked 44th in the 2017 US News Best National Universities.

迈阿密大学的科研与教学十分有名,办学特色采取小班教学,学校现在设有12个学院,主要专业有:应用海洋物理、建筑学、建筑工程、工业工程、机械工程、艺术史、生物学、管理学、国际贸易、生物医学工程、音乐教育等。其本科教育尤为出色,教学设计非常的人性化,培养出很多名人校友。

UM is celebrated for its scientific research and education, as well as its featured small class teaching. There are 12 colleges/schools, in this university including major disciplines of Applied Maritime Physics, Architecture, Civil Engineering, Industrial Engineering, Art History, Biology, Management, International Trade, Biomedicine Engineering, Music Education, etc. Its undergraduate education stands out for its humanistic pedagogy program, which has cultivated many famous alumni.

迈阿密大学的校园环境和地理位置均十分优越,校园风景如画、环境极佳,校园中心有一处与外界连通的景观湖,到处分布着茂盛的植物,被誉为全美校园环境最优美的大学之一。同时,学校所处的地区是迈阿密历史最为悠久的地区之一,这里有着浓郁的欧洲风情和深厚的历史底蕴,颇具特色。

UM boasts an ideal campus and location where people can enjoy a picturesque landscape and pleasurable environment. In the middle of the campus there is a scenic lake that connects to the outside, with luxuriant vegetation everywhere. All these have won UM a great reputation as one of the most beautiful campuses in the US. At the same time, UM is situated in one of the most historically renowned district of Miami, celebrated for its characteristic European style and time-honored atmosphere.

团队简介
Team Profile

指导教师
Instructors

Allan Shulman

美国迈阿密大学建筑学院教授。在过去的十七年中，舒尔曼教授利用迈阿密和迈阿密海滩的城市，探索了二十世纪城市文化与建筑与迈阿密及其海洋城市之间的相互关系。作为一个学者，他以热带建筑和热带文化理念等现代城市丰富的物质基础为研究领域，研究城市转型。他的学术活动还包括建筑展览、设计竞赛、专家研讨会，讲座和小型研讨会，所有这一切都是为了扩大对南佛罗里达州的理解与研究。舒尔曼教授在 1995 年创立了 Shulman+Associates 建筑事务所。

Professor at School of Architecture, University of Miami. In the past 17 years, Prof. Shulman has explored the interrelationship between 21st century urban culture and architecture with Miami and its coastal cities. As a scholar, he is interested in research based on such rich tangible aspects of the modern cities as tropical architecture and tropical cultural concepts, and devotes himself to research on urban transition. His academic activities also include architecture exhibitions, design competitions, experts seminar, lectures and small discussions, all aimed for a wider understanding of southern Florida. Prof. Shulman founded his practice of Shulman + Associates in 1995.

裴钊
Pei Zhao

美国迈阿密大学建筑学院客座教授。研究集中在拉美现代建筑历史与理论研究，中国近现代建筑理论与现代化进程研究，以及基础设施都市化研究。在从事建筑教育之前，拥有 10 年在北美和中国地区的从业经验，既有偏重于城市与建筑设计领域工作经验，也有偏重大型综合项目咨询与管理的经验，以及多家国际性外企管理经验。曾主持和参与了很多综合性、跨领域、不同尺度的城市与建筑设计项目，其中多个获得了不同国家与地区的奖项。

Guest Professor at School of Architecture, University of Miami. His research is concentrated on Latin American architectural history and theory, Chinese modern and contemporary architectural theory and progress of modernization, and the urbanization of infrastructure. Before undertaking architectural education, he has gained 10 years of practical experience in North America and China – not only in terms of urban and architectural design but consultation and management of large-scale comprehensive projects, as well as management of multiple international enterprises. He has conducted and contributed to numerous comprehensive and interdisciplinary urban and architectural design projects of various scales, many of which have been awarded in different countries and regions.

李春青
Li Chunqing

北京林业大学城市规划与设计博士，北京建筑大学建筑学院副教授、硕士生导师，北京未来城市设计高精尖创新中心办公室副主任。主要从事建筑设计理论与教育研究、中国传统园林的现代传承、建筑遗产保护与利用。先后赴德国柏林工业大学、英国格莱摩根大学、美国奥本大学等高校交流访学。近年来主持和参与多项国家级和省部级课题，参与的科研成果获得 3 项"精瑞科学家技术奖"等；教学注重中国传统文化与现代设计教育结合，获得全国高等学校建筑专业指导委员会组织的建筑设计教学成果"优秀教案"奖，指导学生多次获得建筑设计竞赛奖项。

PhD in urban planning and design at Beijing Forestry University; Associate professor and master supervisor at College of Architecture and Urban Planning, Beijing University of Civil Engineering and Architecture; deputy office director at Beijing Advanced Innovation Center for Future Urban Design. She is dedicated to architectural design theory and education research, modern interpretation of traditional Chinese gardens, and conservation and development of built heritage. She has visited Technical University of Berlin, University of Glamorgan, and Auburn University on scholarly trips. In recent years, she has conducted and contributed to various national and provincial/ministry-level research projects, and her achievements has been awarded three Jingrui Technology Awards for Outstanding Scientists. Her pedagogy emphasizes the combination of traditional Chinese culture and modern design education, which has been awarded "Teaching Plan of Excellence" of Architectural Design Teaching Results organized by Architecture Advisory Board of National Institutions of Higher Learning. Students under her supervision have won multiple architectural design competition awards.

王如欣
Wang Ruxin

意大利都灵理工大学建筑设计博士，北京建筑大学建筑学院城乡规划系讲师。主要从事城市设计、文化遗产保护和城市棕地更新方面的研究。作为城乡规划系教师，主要承担本科二年级和四年级城市规划与建筑设计教学工作。

PhD in architectural design at Polytechnic University of Turin, lecturer at Urban-rural Planning Department, College of Architecture and Urban Planning, Beijing University of Civil Engineering and Architecture. She focuses on urban design theory, cultural heritage conservation and urban brown field renewal. As a lecturer at Urban-rural Planning Department, she is in charge of teaching urban planning and architecture design programs for undergraduates of second and fourth year.

团队学生
Team students

岳天琦

建筑学 - 研 2017 级

Yue Tianqi
Architecture - Postgrad 2017

于遨坤

建筑学 - 研 2017 级

Yu Aokun
Architecture - Postgrad 2017

程鹏

建筑学 - 研 2017 级

Cheng Peng
Architecture - Postgrad 2017

郗嘉琦

建筑学 - 研 2017 级

Xi Jiaqi
Architecture - Postgrad 2017

张玮靓

风景园林学 - 研 2017 级

Zhang Weiliang
Landscape Architecture - Postgrad 2017

王欣雨

城乡规划学 - 研 2016 级

Wang Xinyu
Urban and Rural Planning - Postgrad 2016

孙小鹏

工业设计工程 - 研 2016 级

Sun Xiaopeng
Industrial Design Engineering - Postgrad 2016

翟玉琨

建筑学 - 研 2015 级

Zhai Yukun
Architecture - Postgrad 2015

韩翘楚

建筑学 - 本 2013 级

Han Qiaochu
Architecture - Undergraduate 2013

王玉珏

建筑学 - 本 2015 级

Wang Yujue
Architecture - Undergrad 2015

杜一凡

城乡规划学 - 本 2015 级

Du Yifan
Urban and Rural Planning - Undergrad 2015

王颢翔

城市道路与桥梁工程 - 本 2015 级

Wang Haoxiang
Urban Road and Bridge Engineering - Undergrad 2015

工作照
Teams at work

工作营任务书
Workshop Assignment

项目背景介绍
Project background introduction

迈阿密海滩的北海滩区，是一处被海水包围、仅高出海平面1米的海滨区。这个城区有3万居民和无数游客，称为北海滩区，是迈阿密海滩的第2个市中心区，两侧分别是大西洋和比斯坎湾。它是在南海滩开发后发展起来的。南海滩不仅更有名，而且从20世纪末到21世纪极度繁荣，而北海滩则遭到大家的忽视，并有撤资。其中的一系列海滨空地是恢复其活力的关键。

Miami beach's North Shore, a coastal strand surrounded by water and rising little more than a meter above sea - level, is home to more than 30,000 residents and countless more tourists. The urban district, called North Beach, is the second urban center of Miami Beach, facing the Atlantic Ocean on one side and Biscayne Bay on the other; it was developed a generation after the better - known resort and residential districts of South Beach. While South Beach has prospered, experiencing extraordinary development in the late 20th and 21st c, North Beach has suffered neglect and disinvestment. A series of open lots along its oceanfront are considered one key to its revitalization.

这个区域的建筑类型、尺度和风格有很强的一致性。城市结构是紧凑、较低和中型尺度的建筑，以度假为主，混合着小型商业中心，提供了极有吸引力的居民生活方式。该地区的大部分地块被列入《美国国家历史地名录》，当地现有建筑即将列入当地的历史建筑名录。

The area is marked by a strong congruence of building type, scale, and architectural styles. The urban structure of compact, low and mid - scale architecture, which are mainly resort architectures, mixed with small commercial centers and main streets, provides an attractive lifestyle. Most of the district is currently listed on the National Register of Historic Places, and local historic architecture designation of existing building stock is imminent.

设计任务及要求
Design tasks and requirements

该项目基地的特性引入了在总体规划和建筑设计中要考虑的几个主题:
The project site introduces several topics to be considered in both master planning and building design:

- 制定一个城市设计框架,创造性地用现有的基础设施为公众提供便利。
- *Develop an urban design framework that provides public amenity and deals creatively with existing infrastructure.*
- 构想一个具有当代功能的都市度假村。
- *Conceive a contemporary program for an urban resort city.*
- 平衡功能需求,例如在度假村和住宅用途之间,以及停车场和可居住空间之间进行平衡。
- *Balance programmatic needs, for instance between resort and residential uses, and between parking and habitable space.*
- 在实现总体设计目标时,创造性地处理建筑类型的问题。
- *Work creatively with issues of building typology in achieving overall design objectives.*
- 展示沿海建筑设计的弹性模式。
- *Demonstrate resilient models of coastal construction.*

方案过程需要针对海滨开发、法规和用地的当前模式建立批判性的方法,从而把控当前项目的规划和设计区域。

A critical approach toward current models of oceanfront development, regulations, land use is required to master plan and design areas of the current project.

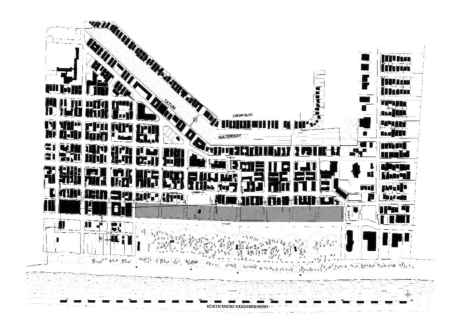

项目信息
Project information

1. 选址

1. Sites

选择的地段是北海滩区北缘未经设计的街道立面，对面是北海滩露天公园。这8个街区（每个面积约5,100平方米）如今几乎都已闲置，主要用于停车。然而，它们曾经也是属于一片繁荣的度假区。

The sites chosen present a street facade without design on the northern edge of the North Beach district, facing North Shore Open Space Park. The eight blocks (each approx. 5,100m2), are almost empty today, and used mainly for parking. However, at one time they were part of a thriving resort district.

最初在1919年规划的海上高地街区中综合设置了酒店和住宅。从20世纪50年代到80年代，迈阿密海滩市买下了柯林斯大道两侧的土地，并组合起来建造新公园。 1989年，该市将所有的土地都转让给了佛罗里达州。政府拆除了柯林斯大道东侧的酒店和住宅，目的是要创建一个区域性海滩公园和开放式空间保护区——即现在的北海滩露天公园。柯林斯大道西侧的地段、沿海公路、柯林斯大道都用于公园的停车。最近，该州将这些土地的所有权又转让回了迈阿密海滩市。

First platted in 1919, the Altos del Mar (Highlands of the Sea) neighborhood counted a mix of hotels and home sites. Between the 1950s and 80s, the City of Miami Beach purchased and assembled the land on both sides of Collins Avenue for the purpose of constructing the new park. In 1989, the City deeded all of the land to the State of Florida. Hotels and homes on the east side of Collins Avenue were removed by the government to create a regional beach park and open - space preserve - North Shore Open Space Park. The lots on the west side of Collins Avenue, the coastal highway, Collins Avenue, were designated as parking for the park. More recently, the State deeded the parcels back to the City of Miami Beach, which continues to own the land.

今天，这个28英亩的公园与北海滩的城市肌理完全断开，几近荒废。然而，总部位于鹿特丹的West 8 建筑设计事务所正在进行新的总体规划，以重新活化公园。柯林斯大街西侧的停车场基本闲置。单独来看，项目地段的8个街区拥有了一个填充式开发的机会。然而从总体上看，这又是一个独特的机遇，人们将在这个十分罕见的海滨公园地段中测试和推进城市发展的新模式。未来的再开发规划需要考虑邻近社区的需求和尺度，一部分要建设商业、住宅和度假功能的综合区，另外还应适当地考虑停车空间。

The 28 - acre park, today completely disconnected from the urban fabric of North Beach, is little - used. However, a new masterplan by the Rotterdam - based West 8 Architectural Design Offical to revitalize the park is underway. The parking lots on the west side of Collins Avenue stand mainly empty. Individually, the eight blocks of the project site represent an opportunity for infill development. In aggregate, however, it represents a singular opportunity to test and advance new models of urban development along a rare stretch of oceanfront parkland. Future redevelopment plans need to consider the needs and scale of the adjacent neighborhood, partly by proposing a careful mix of commercial, residential and resort programs. Appropriate provision for parking will need to be considered.

2. 海平面上升
2. Sea Level Rise

该地区的海平面上升问题将会成为方案地段规划和设计的重要组成部分。迈阿密海滩近年来潮汐泛滥有所加剧,关于海平面上升模式的争论仍在继续,"东南佛罗里达地区气候变化协议组织"预测:"到2030年,海平面将比1992年测量的平均值上升6至10英寸。到2060年,海平面上升预计将比1992年水平高出14至34英寸。到2100年,它可能会比1992年高出31到81英寸。"

The question of sea level rise will be an important component of both planning and design on the proposed sites. Miami Beach has experienced increased tidal flooding in recent years. The debate over sea - level rise patterns continues, however the Southeast Florida Regional Climate Change Compact has predicted "that the sea level would rise between 6 to 10 inches by 2030 from the mean sea level measured back in 1992. By 2060, sea - level rise is projected to be at 14 to 34 inches above 1992 levels. By 2100, it could be 31 to 81 inches higher than what it was in 1992."

迈阿密海滩市目前正在制定应对偶发性洪水的对策,如将道路提高约45厘米,并安装水泵给受淹区域排洪。该市还建议修改迈阿密海滩市区区划条例,促进产生能应对气候变化和海平面上升的优秀规划实践案例。

The City of Miami Beach is currently responding to episodic flooding by raising roadways about 18 inches (45cm), and by installing pumps to drain flooded areas. It has also proposed revising the City of Miami Beach Zoning Ordinance to incentivize good practices with regard to climate change and sea level rise.

工作营日程
Workshop Schedule

时间 Time		任务 Task
第一天 Day 1	上午 Morning	参观迈阿密大学 Visit university of Miami 工作营开营 Set up studio 迈阿密海滩讲座/项目介绍 Lecture on Miami Beach/Introduction to project area
	下午 Afternoon	迈阿密海滩讲座/项目介绍 Lecture on Miami Beach/Introduction to project area
第二天 Day 2	上午 Morning	南海岸调研 Survey South Beach
	下午 Afternoon	会见迈阿密海滩城市设计总监汤姆·穆尼 Meeting with Tom Mooney, Planning Director City of Miami Beach 参观北海岸/参观项目基地 Visit to North Beach/visit to site
第三天 Day 3	上午 Morning	调研布凯尔市区的帕姆艺术博物馆并了解霜冻学 Brickell/Downtown PAMM Art Museum & Frost Science
	下午 Afternoon	怀恩伍德/设计区调研 Survey Wynwood/Design District 参观Shulman + Associates建筑事务所 Visit Shulman + Associates
第四天 Day 4	上午 Morning	工作环节 Working Session
	下午 Afternoon	
第五天 Day 5	上午 Morning	区域总体规划讲习班 Area Master Plan workshop
	下午 Afternoon	
第六天 Day 6	上午 Morning	工作环节 Working Session
	下午 Afternoon	工作营总计划确定 Master Plan Pin - up
第七~九天 Day 7 ~ 9	上午 Morning	工作环节 Working Session
	下午 Afternoon	

团队成果简介与展示
Team Achievements and Presentation

组一：

Group 1:

本案设计将切入点放在加强基地的竖向联系上，用架步行桥的方式解决城市主路打断基地与海滩联系的现状问题，并在几个街区间的竖向道路上布置商业和公寓，每个街区内形成服务性质的公共空间，增强基地活力。

The design starts from strengthening the vertical connections of the site. With a pedestrian bridge, it attempts to solve the existing issue of an urban artery interrupting the connection between the site and the beach. Commercial and residential buildings are planned to be distributed on the vertical roads among the blocks, each with its public space for service purposes, thus invigorating the site.

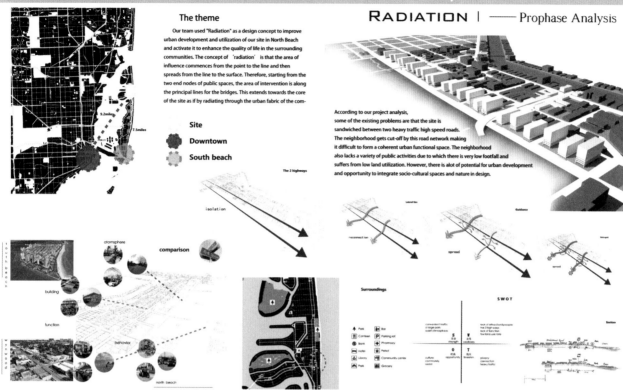
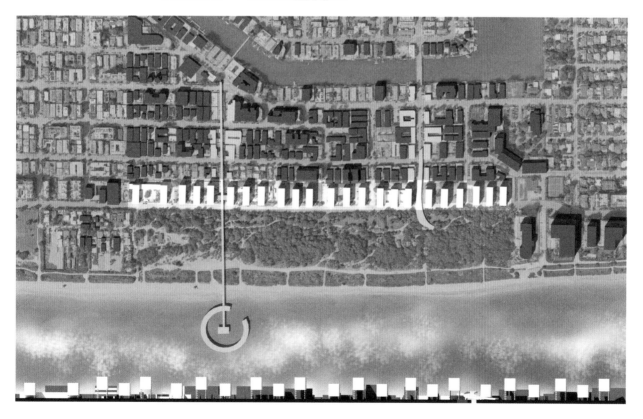

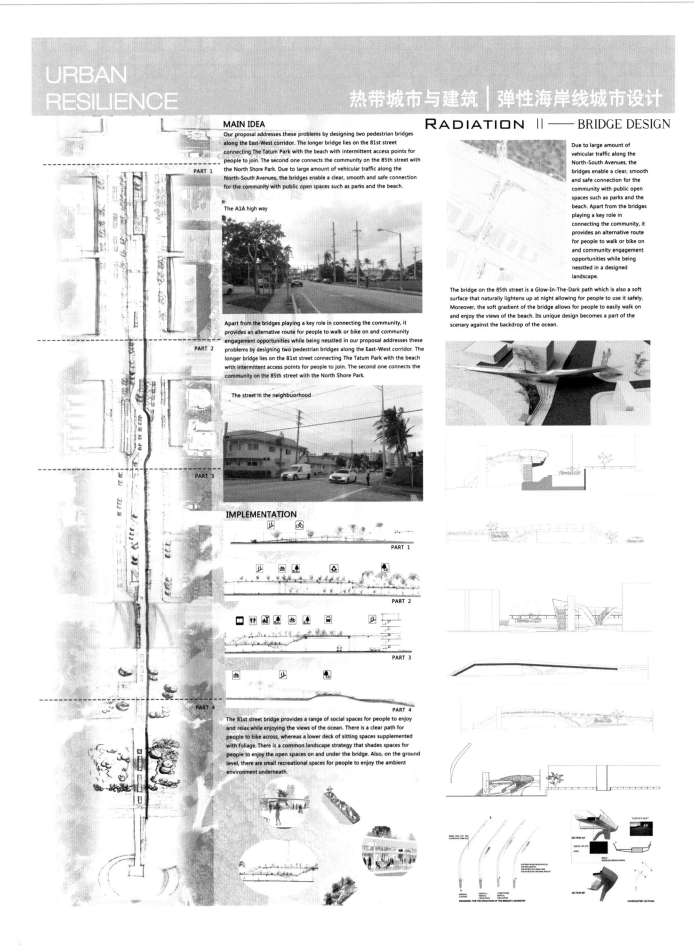

URBAN RESILIENCE

RADIATION Ⅲ —— BLOCK DESIGN

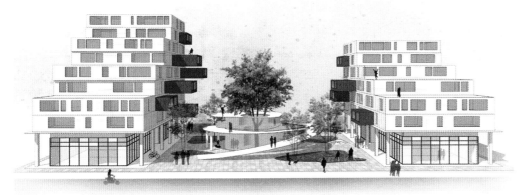

Situation problems
The neighborhood lacks a variety of public activities due to which there is very low footfall and suffers from low land utilization.
Hard to see the beach is another problem, although there is near by the sea.

Strategy analysis
According to the previous analysis, the crosswise expressway will interrupt the flow of people, so we find two longitudinal axes to stress the longitudinal relationship of the area in order to find a more suitable space for pedestrians

Two apartments are arranged along the streets of each block. New functions are implanted into the block and public buildings are arranged in the middle. The layout of the building gradually made people flow into every block.

The ladder and the extended box can make the households have a better view.

Master Plan

COLLINS AVENUE

We hope to focus our attention on community environment creation, create a dynamic community and enrich the community life of local residents. As we have analyzed in front of us, we design two longitudinal walking bridges, strengthen the longitudinal path between 8 blocks, and gradually infiltrate the stream from east to west into land, park and beach. In each block, there are two high-rise apartment buildings along the street, one part of which is commercial retail, forming a business atmosphere and passing through the space for people. Public buildings are arranged in the middle, such as libraries, cafes and galleries, forming open community public spaces, carrying the entertainment life of nearby residents. Building monomer is stepped and block interspersed, fully meeting the needs of households to see the sea.

Floors Plan

Section

Multiple combinations

Garden | Hall | Court | Building+Coverway

Building is open to the street. | Yard is open to the street. | Platform with roof | Two different yards

组二：

Group 2:

本案设计将场地前快速路架高，形成具有雕塑感的立体交通。这使场地与滨海公园完全打通，城市到沙滩的障碍被消除。利用住宅底层与桥下空间形成商业和艺术混合群，服务和吸引人流，形成具有艺术氛围的滨海活动与居住区。

The design is noted for its elevation of the expressway in front of the site, creating a sculpture-like three dimensional traffic. This fully opens the site to the seaside park, eliminating the barriers between the city and the beach. A mixed commercial and art complex is created from the ground floor of the residential buildings and the space under the bridge, which both serves and attracts people, generating a coastal leisure and residential area with an artistic touch.

URBAN RESILIENCE

热带城市与建筑 | 弹性海岸线城市设计

流动 FLUX

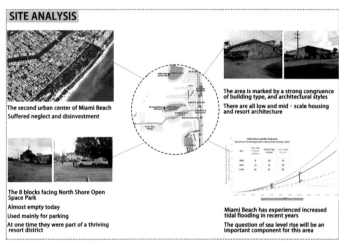

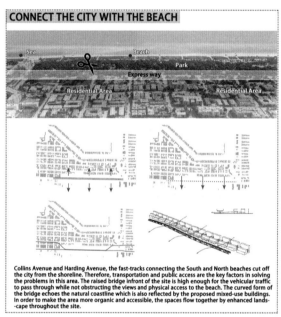

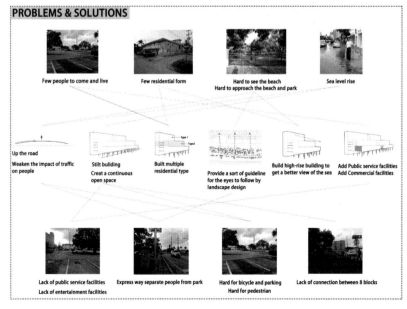

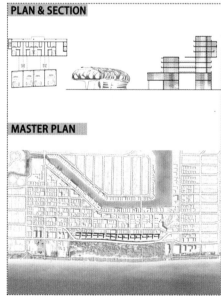

URBAN RESILIENCE

热带城市与建筑 | 弹性海岸线城市设计

流动 FLUX

Miami Beach 充满阳光和活力，成为一片休闲、娱乐和度假的圣地。这里吸引了许多人来此定居，也吸引了世界各地人们来此旅游度假。由此带来了大量的城市建设，以满足人们生活和娱乐需求。但我们在南海滩和北海滩看到了两种不同的发展情况。一个是极其繁华热闹的商业旅游区，一个是安静的居住区。

The project explores the possibility of solving the urban problem on a comprehensive way. A new ground plaza is created to reactivate the site; a series of public spaces are designed according to the circulation of pedestrians; new public buildings and housing are added for the rehabilitation of urban fabric; Negative parking lot is turn into public space for picnic, sports and shopping; New programs like library, shopping mall, museums are created to turn the negative site into a positive space with the connection between the beach and residential area. By design, the North Beach will be a haven of peace, a health park for relaxation and a vibrant arts venue.

住宅 Residence

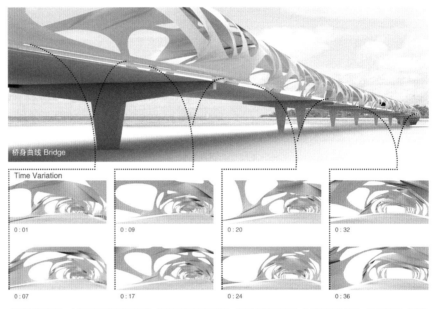
桥身曲线 Bridge

Time Variation
0:01　0:09　0:20　0:32
0:07　0:17　0:24　0:36

为了将北海滩转变为一个适于居住和休闲的城市区域，我们需要提高这里的可达性，打通城市与海滩的连接，让海滩与居住区成为一个有机的整体，让海滩生活成为城市生活的重要组成。所以，交通成为解决这片区域问题的关键因素。Collins Ave 和 Harding Ave 作为连接南海滩和北海滩的快速道路，割裂了城市和海滩。所以，我们将道路架高，让车辆从空中通过。

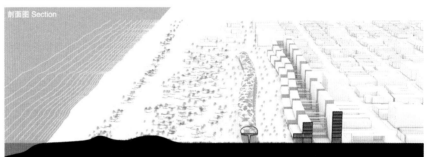
剖面图 Section

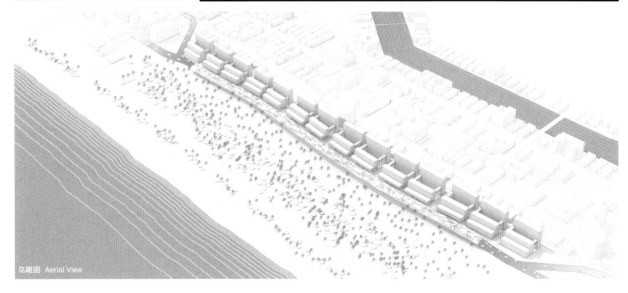
鸟瞰图 Aerial View

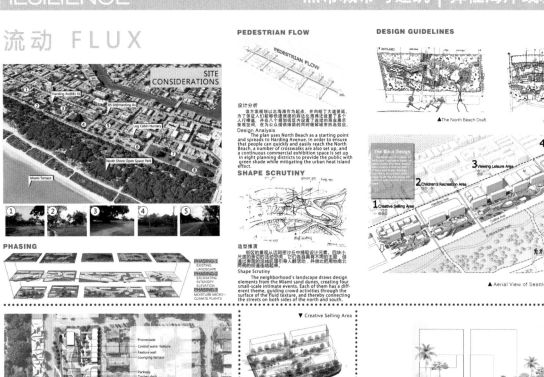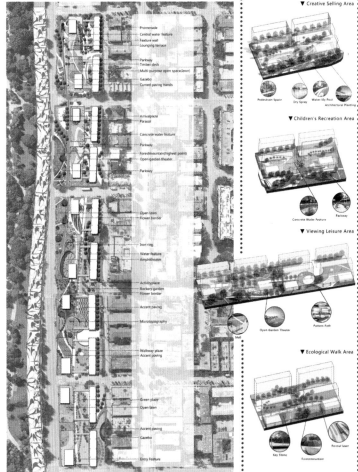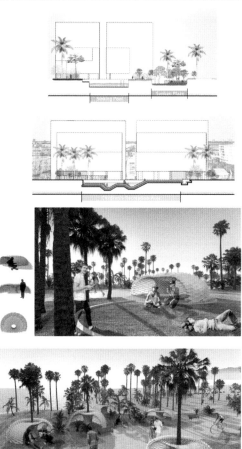

组三：
Group 3:

本案设计主题为"超级覆盖物"，将基地的活动空间整体抬高至二层，提供良好的景观视线的同时也创造了更多不同功能的公共空间，建筑与植物共存，使城市与自然的过渡不太生硬。

The desgin, themed "Mega-Shelter", raises the entire functional space of the site to the second floor, which provides an enjoyable view of the landscape while creating a variety of public spaces with different functionalities. Symbiosis between buildings and vegetation ensures a harmonious transition from urban to nature.

URBAN RESILIENCE

热带城市与建筑 | 弹性海岸线城市设计

INVESTIGATION AND RECONNAISSANCE

Pension　　Bustling　　Quiet　　Desolate　　Tropical Plants　　Concise　　Courtyard　　Multifunction

Miami is an ecologically rich city replete with many beautiful tropical landscapes. However, the existing landscape is mostly ornamental as people are not able to interact with it. Also, the city's living and ecological spaces are not well integrated within the urban fabric. Miami Beach is originally a place for locals to enjoy their life leisurly. With growing tourism and bustling commercial spaces, South Beach is recognized as the 'city card'.

Element Collision　　Greening　　Overhead　　Public Art

SITE STATUS INVESTIGATION

  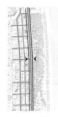

Roadway　　Virescence　　Street-side

North Beach as an area has not been developed to its full potential. There are a lot of residential houses around our site, but there is a lack of centralized living facilities and entertainment venues. The North Shore park is beautifully landscaped, but lacks social engagement activities. Hence, we want to create it as a living space that caters to the social requirements of local residents and visitors.

MANUAL　　　　　　　　　NATURAL

Concept Generation

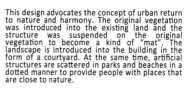

This design advocates the concept of urban return to nature and harmony. The original vegetation was introduced into the existing land and the structure was suspended on the original vegetation to become a kind of "mat". The landscape is introduced into the building in the form of a courtyard. At the same time, artificial structures are scattered in parks and beaches in a dotted manner to provide people with places that are close to nature.

The surface of architectural in Miami is very rich, with punctate, linear, facet and irregular shapes. Building skin makes good use of local materials to create special texture. The bold use of colors in architecture reflects the aesthetic of locals.

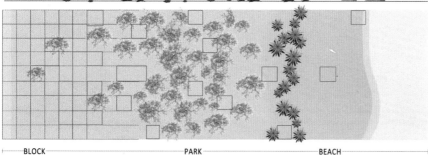

BLOCK　　　　PARK　　　　BEACH

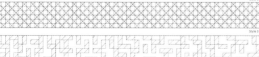
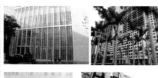
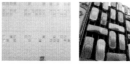

URBAN RESILIENCE

热带城市与建筑 | 弹性海岸线城市设计

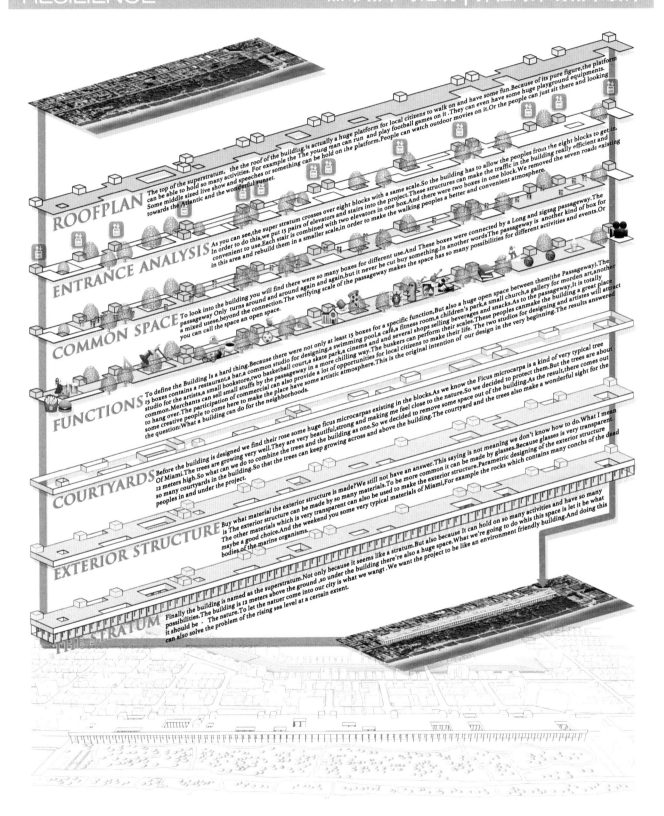

URBAN RESILIENCE

热带城市与建筑 | 弹性海岸线城市设计

NORTH BEACH POSITION ANALYSIS

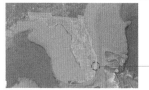

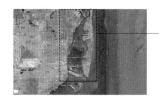
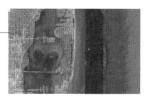

- Florida is located in the southeastern United States, near Kansas. The latitude of Florida is very low, it is a good place for people to spend a holiday.
- Miami is located on the southern tip of Florida. Miami is the most developed city in Florida, close to Everglades Park.
- Miami Beach is located on a small island on the eastern side of Miami. Miami Beach is densely populated, commercially prosperous and has many attractions.
  - The small island in the middle is a rich area (green area).
  - The big island on the east side is the territory of the people (orange area).
- North and south beaches are located on both sides of the island. Miami Beach is densely populated, commercially prosperous and has many attractions. South Beach is always full of noisy. We may take a quiet place for the elderly and young people to come here for a holiday. Maybe North Beach is a good choice!

HISTORICAL EVOLUTION

1896	1920s	1930s	1959	1970s	1980s	1990s	NOW
The beginning of the city	The immigrants moved from the north.	The Development because of the Art Deco. The resort of the rich	Immigration wave from Cuba	Economic prosperity	Vice City Full of Drug Violence Crime	Prosperous of Tourism and Consumption	"A vacation spot for retirees"

BEACH AREA GREE DESIGN

OVERLAY DIFFERENT LAYERS

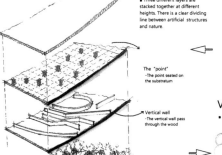

- Three different layers are stacked together at different heights. There is a clear dividing line between artificial structures and nature.
- The "point" — The point seated on the substratum
- Vertical wall — The vertical wall pass through the wood
- Substratum

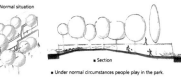

Plan

- The beach area is covered by green plants. Vertical wall are shaded between green plants, red artificial structures "point" are distributed on the grid's focal point

VERTICAL WALL

- Normal situation

Section
- Under normal circumstances people play in the park.

- Sea level rise

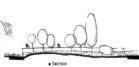

Section
- When the sea level rises, people do activities on vertical walls.

ARTIFICIAL STRUCTURE "POINT"

- We abstract the structure into three layers: the upper, middle, and lower layers. The cubes in each layer can be evolved into various constructions. For example, the upper layer changes to a different type of roof structure, the middle layer changes to different forms of stairs, and the lower layer changes to various types. supporting structure.

- The upper layers

A1 CURVE   A2 FOLDED PLATE   A3 FLAT

- Middle layers

B1 SPIRAL STAR   B2 CURVE STAR   B3 FLAT

- lower layers

C1 SUSPENDED STRUCTURE   C2 STEEL FRAME STRUCTURE   C3 BEAM STRUCTURE

- We placed the most active point, a structure composed of three layers of active parts, in the orange area, and we look forward to providing visitors with a lot of activity while at the same time inspiring them. In the same way, the structure with the quietest part is placed in the blue area of the drawing, and the structure with good dynamics and quietness is placed in the red and blue areas in the area.

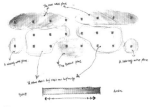

- There are three forms of change at each of the three levels, so there are a total of 27 combinations

For example

- TYPE1: A1+B1+C1 (The most active one)
- TYPE10: A1+B1+C2
- TYPE14: A2+B2+C2
- TYPE21: A3+B1+C3

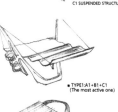

- TYPE15: A3+B2+C2
- TYPE8: A2+B3+C1
- TYPE18: A3+B2+C2 (The most quitness one)
- TYPE25: A1+B3+C3